Quantum Field Theory

A quantum computation approach

Quantum Field Theory

A quantum computation approach

Yannick Meurice

*Department of Physics and Astronomy, The University of Iowa,
Van Allen Hall, Iowa City, IA 52242, USA*

IOP Publishing, Bristol, UK

ISBN   978-0-7503-2187-7 (ebook)
ISBN   978-0-7503-2185-3 (print)
ISBN   978-0-7503-2188-4 (myPrint)
ISBN   978-0-7503-2186-0 (mobi)

DOI   10.1088/978-0-7503-2187-7

Version: 20210201

IOP ebooks

British Library Cataloguing-in-Publication Data: A catalogue record for this book is available from the British Library.

Published by IOP Publishing, wholly owned by The Institute of Physics, London

IOP Publishing, Temple Circus, Temple Way, Bristol, BS1 6HG, UK

US Office: IOP Publishing, Inc., 190 North Independence Mall West, Suite 601, Philadelphia, PA 19106, USA

Cover image. The cover picture is from Professor Norbert Linke's lab at the Joint Quantum Institute, University of Maryland, Department of Physics. His experimental group is working with trapped atomic ions on different applications in quantum physics. Our groups are now collaborating on a project on real-time evolution and measurements of phase shifts. The photo was taken by Norbert Linke and edited by Nova Meurice.

Dedicated to Hallsie, Marielle, Nova, Naomi and the memory of Emma and Jeannine.

Contents

Preface		**xi**
Acknowledgements		**xii**
Author biography		**xiii**
1	**Introduction**	**1-1**
1.1	Goals of the lecture notes	1-1
1.2	Classical electrodynamics and its symmetries	1-2
1.3	Field quantization	1-2
1.4	The need for discreteness in quantum computing	1-3
1.5	Symmetries and predictive models	1-4
	References	1-5
2	**Classical field theory**	**2-1**
2.1	Classical action, equations of motion and symmetries	2-1
2.2	Transition to field theory	2-4
2.3	Symmetries	2-5
2.4	The Klein–Gordon field	2-9
2.5	The Dirac field	2-11
2.6	Maxwell fields	2-17
2.7	Yang–Mills fields	2-19
2.8	Linear sigma models	2-20
2.9	General relativity	2-22
2.10	Examples of two-dimensional curved spaces	2-27
2.11	Mathematica notebook for geodesics	2-32
	References	2-37
3	**Canonical quantization**	**3-1**
3.1	A one-dimensional harmonic crystal	3-1
3.2	The infinite volume and continuum limits	3-6
3.3	Free KG and Dirac quantum fields in $3+1$ dimensions	3-10
3.4	The Hamiltonian formalism for Maxwell's gauge fields	3-13
4	**A practical introduction to perturbative quantization**	**4-1**
4.1	Overview	4-1
4.2	Dyson's chronological series	4-2

4.3   Feynman propagators, Wick's theorem and Feynman rules   4-4
4.4   Decay rates and cross sections   4-6
4.5   Radiative corrections and the renormalization program   4-7
      References   4-7

5   The path integral   **5-1**

5.1   Overview   5-1
5.2   Free particle in quantum mechanics   5-3
5.3   Complex Gaussian integrals and Euclidean time   5-4
5.4   The Trotter product formula   5-10
5.5   Models with quadratic potentials   5-12
5.6   Generalization to field theory   5-14
5.7   Functional methods for interactions and perturbation theory   5-15
5.8   Maxwell's fields at Euclidean time   5-16
5.9   Connection to statistical mechanics   5-16
5.10 Simple exercises on random numbers and importance sampling   5-18
5.11 Classical versus quantum   5-33
      References   5-34

6   Lattice quantization of spin and gauge models   **6-1**

6.1   Lattice models   6-1
6.2   Spin models   6-2
6.3   Complex generalizations and local gauge invariance   6-4
6.4   Pure gauge theories   6-5
6.5   Abelian gauge models   6-6
6.6   Fermions and the Schwinger model   6-7
      References   6-8

7   Tensorial formulations   **7-1**

7.1   Remarks about the discreteness of tensor formulations   7-1
7.2   The Ising model   7-3
7.3 $O(2)$ spin models   7-5
7.4   Boundary conditions   7-7
7.5   Abelian gauge theories   7-7
7.6   The compact abelian Higgs model   7-10
7.7   Models with non-abelian symmetries   7-11
7.8   Fermions   7-13
      References   7-13

8	**Conservation laws in tensor formulations**	**8-1**
8.1	Basic identity for symmetries in lattice models	8-1
8.2	The $O(2)$ model and models with abelian symmetries	8-2
8.3	Non-abelian global symmetries	8-4
8.4	Local abelian symmetries	8-6
8.5	Generalization of Noether's theorem	8-7
	References	8-8

9	**Transfer matrix and Hamiltonian**	**9-1**
9.1	Transfer matrix for spin models	9-1
9.2	Gauge theories	9-5
9.3	$U(1)$ pure gauge theory	9-8
9.4	Historical aspects of quantum and classical tensor networks	9-10
9.5	From transfer matrix functions to quantum circuits	9-12
9.6	Real time evolution for the quantum ising model	9-14
9.7	Rigorous and empirical Trotter bounds	9-19
9.8	Optimal Trotter error	9-22
	References	9-28

10	**Recent progress in quantum computation/simulation for field theory**	**10-1**
10.1	Analog simulations with cold atoms	10-1
10.2	Experimental measurement of the entanglement entropy	10-5
10.3	Implementation of the abelian Higgs model	10-7
10.4	A two-leg ladder as an idealized quantum computer	10-9
10.5	Quantum computers	10-11
	References	10-13

11	**The renormalization group method**	**11-1**
11.1	Basic ideas and historical perspective	11-1
11.2	Coarse graining and blocking	11-2
11.3	The Niemeijer–van Leeuwen equation	11-5
11.4	Tensor renormalization group (TRG)	11-8
11.5	Critical exponents and finite-size scaling	11-11
11.6	A simple numerical example with two states	11-13
11.7	Numerical implementations	11-19

11.8 Python code   11-20
11.9 Additional material   11-30
     References   11-31

12   Advanced topics   **12-1**

12.1 Lattice equations of motion   12-1
12.2 A first look at topological solutions on the lattice   12-4
12.3 Topology of $U(1)$ gauge theory and topological susceptibility   12-7
12.4 Mathematica notebooks   12-15
12.5 Large field effects in perturbation theory   12-27
12.6 Remarks about the strong coupling expansion   12-30
     References   12-32

Appendices

Appendix A   **A-1**

Appendix B   **B-1**

Preface

The idea that things are made out of particles is very old. However, at the time of Ludwig Boltzmann's death it was still very controversial. The twentieth century was marked by the discoveries of many particles, together with the evidence for a completely new type of behavior in nature. The laws of quantum mechanics are more bizarre than the wildest pre-Socratic dreams, but if we look closely enough they explain how things work.

Quantum field theory was developed at the same time as particle physics. Its self-consistency led to predictions of the the existence of particles which were later discovered experimentally. The electro-weak sector can be understood very well with perturbation theory and Feynman diagrams. However, the strong interactions require new methods. The use of a space–time lattice allows physicists to use statistical sampling to calculate accurately the masses and form factors of hadrons.

Calculating the real-time unitary evolution and finite-density behavior of strongly interacting quantum systems are major challenges for the twenty-first century. They require computational resources far beyond the capabilities of classical super-computers. The idea that quantum devices could be used to deal with these quantum problems has been established in principle for systems with quasi-local interactions. We are now at the beginning of a new and exciting era where we get to put these general ideas into practical forms.

Recent experimental progress in atomic, molecular and condensed matter physics have made possible sophisticated manipulations of small Hilbert spaces. We can use these new techniques to perform computations related to quantum models. This could revolutionize our understanding of particle collisions, nuclei, early cosmology and black holes, not to mention practical problems.

The goal of this book is to connect the basic aspects of lattice field theory to their continuum versions and to help invent new methods to use quantum computers or quantum simulation experiments to perform real-time calculations for strongly interacting systems. I believe that technological progress and creative algorithmic ideas will allow us to do real-time calculations in quantum chromodynamics in the next decade. It may take longer. The advent of quantum computing is as ineluctable as Joyce's modality of the visible.

I hope the next generations of physicists will enjoy these successes and that they will find the book useful. It is said that Thales of Miletus fell in a well while looking at the stars. It is not always easy to work for the future.

This book would not have been possible without the continued support of my family, students, colleagues, grant monitors, neighbors and soccer friends in Iowa City.

Acknowledgements

I thank my graduate students who played a major role in the development of tensor methods discussed in the book and especially Yuzhi Liu, Haiyuan Zou, and Judah Unmuth-Yockey. I would like to thank the late D Speiser, J Weyers and C Itzykson for their teaching on group theory, Pontryagin duality, Peter-Weyl theorem and strong coupling expansions. I learned a lot from conversations with Mari-Carmen Banuls, Alexei Bazavov, David Berenstein, Immanuel Bloch, Richard Brower, Simon Catterall, Shailesh Chandrasekharan, Xi Dong, Stephen Jordan, Seth Lloyd, Michael McGuigan, Lode Pollet, Nikolay Prokofiev, Ryo Sakai, Boris Svistunov, Shan-Wen Tsai, Tao Xiang, Li-Ping Yang, Zhiyuan Xie, Johannes Zeiher, Jin Zhang, and other members of the QuLAT collaboration. I thank Erik Gustafson, Zheyue Hang, Robert Maxton, Daniel Simons, and especially Michael Hite and Naomi Meurice for comments on the manuscript. My colleagues Vincent Rodgers, Hallsie Reno, Wayne Polyzou, Gerald Payne, Bill Klink, Fred Skiff, Phil Kaaret and Robert Merlino have been a major support of my research and teaching for the last thirty years. This work was supported in part by the U.S. Department of Energy (DOE) under Award Numbers DE-SC0010113, and DE-SC0019139. I thank the grant monitors for their continued support.

Author biography

Yannick Meurice

 **Yannick Meurice** is a Professor at the University of Iowa. He obtained his PhD at U.C. Louvain-la-Neuve in 1985 under the supervision of Jacques Weyers and Gabriele Veneziano. He was a postdoc at CERN and Argonne National Laboratory and a visiting professor at CINVESTAV in Mexico City. He joined the faculty of the Department of Physics and Astronomy at the University of Iowa in 1990. His current work includes lattice gauge theory, tensor renormalization group methods, near conformal gauge theories, critical machine learning, quantum simulations with cold atoms and quantum computing. He is the PI of a multi-institutional DOE HEP QuantISED grant.

Quantum Field Theory
A quantum computation approach
Yannick Meurice

Chapter 1

Introduction

1.1 Goals of the lecture notes

The idea that we should use quantum devices to perform computations for quantum problems involving many degrees of freedom, expressed among others by Richard Feynman [1], has received a lot of attention in recent years. In an influential article, Seth Lloyd [2] states that: 'Feynman's 1982 conjecture, that quantum computers can be programmed to simulate any local quantum system, is shown to be correct'. Part of these lecture notes will be devoted to explaining the concrete meaning of this statement in the context of quantum field theory.

Actual physical systems involving cold atoms (see [3] for a review of the early day progress) or trapped ions (see [4] for a recent example) have been used to mimic the behavior of simplified many-body models such as various types of spin chains or Hubbard models. Companies like IBM (https://www.research.ibm.com/ibm-q/) or Rigetti (see https://www.rigetti.com/forest) have made some of their quantum computers available to the general public. It is expected that in the coming years we will be able to develop new methods to use these new tools to perform calculations for quantum field theory (QFT) models that are physically relevant and impossible to perform with ordinary classical computers. In 2019, the QFT models that can be studied with noisy quantum computers (see [5–7] for recent examples) are very simple compared to the models used to do particle physics phenomenology or describe superconductivity. Beginning graduate students interested in contributing to the quantum computing effort need to focus on some aspects of simple QFT models that are not always emphasized in standard textbooks. This is why these lecture notes have been written. Our main goals are:

(i) Give an overview of the logical structure of QFT and describe how to use Feynman rules to calculate decay rates and cross sections.

(ii) Provide an introduction to the use of quantum computing for QFT problems with the practical tools that are available to us today or in the near future.

Learning QFT is a lengthy process and it usually takes the equivalent of two or three semesters of study to be able to do original calculations relevant to the standard model of particle physics. As there are now excellent textbooks devoted to this task, we will only give an overview of the accomplishments and a summary of the Feynman rules relevant for elementary processes. The missing steps in the presentation can be found for instance in the textbook of Peskin and Schroeder [8]. Later we use the abbreviation PS for this reference. In order to facilitate the connection, we will use their notations for vectors, spinors, metric and field creators/annihilators.

1.2 Classical electrodynamics and its symmetries

At the classical level, fields are continuous functions of the continuous space–time. We have a good physical intuition about the effects of the **E** and **B** electromagnetic fields on charged particles. We can visualize the field lines of the Earth's magnetic field by moving a compass and observing the changes of orientation of the needle. Electric fields can be reconstructed from the acceleration of charged particles put in their presence. At each spatial point **x**, we have the electric field vector **E(x)**. This is an example of vector field. It transforms like an ordinary position vector under global spatial rotations.

The **E** and **B** fields mix under Lorentz transformations. They can be understood as the six components of an antisymmetric tensor $F_{\mu\nu}$ where μ and ν are relativistic space–time four-indices. Maxwell equations can be derived by applying the principle of least action for a relativistic invariant action. Note that the tensor field $F_{\mu\nu}$ is constructed in terms of the four-vector field A_μ. This is discussed in chapter 11 of J D Jackson's textbook [9] and reviewed in section 2.5. A_μ can be transformed by adding the four-gradient of an arbitrary function of the space–time variable without changing the physical values of **E** and **B**. This is an example of *local* symmetry. In contrast, Lorentz transformations are global space–time symmetries.

1.3 Field quantization

In the case of the point particles used in classical mechanics, we know how to convert the classical description of the evolution into a quantum mechanical one. There are two ways to proceed. One is to replace the Poisson bracket by commutators. This is the canonical approach. The other is to express the evolution operator as a sum over classical trajectories. This is the path-integral approach. Can we apply similar procedures for fields? It is shown in PS that both approaches can be used to describe free fields and to treat their interactions using perturbative methods. In the canonical approach and in $D = 4$ (three space and one time dimensions), one postulates commutation relations of the form

$$[\varphi(\mathbf{x},\,t),\,\pi(\mathbf{y},\,t)] = i\hbar\delta^3(\mathbf{x} - \mathbf{y}), \tag{1.1}$$

for the field generically denoted as $\varphi(\mathbf{x})$ and its canonical conjugate $\pi(\mathbf{y})$. In the path-integral approach the main object is a partition function

$$Z = \int [\mathcal{D}\varphi] e^{iS[\varphi]/\hbar}, \tag{1.2}$$

where S is the action used in the classical formulation. They provide the same method to calculate cross sections or decay rates using perturbative series expressed in terms of Feynman diagrams.

There is, however, a significant issue associated with the fact that for fields on a continuous space–time we introduce an infinite number of degrees of freedom. Typically, when we calculate the quantum corrections perturbatively, infinities appear. If these infinities can be removed by redefining a finite number of microscopic couplings in order to subtract all the infinities, we say that the theory is perturbatively renormalizable. This also means that we can calculate and make predictions.

1.4 The need for discreteness in quantum computing

The building blocks of ordinary (classical) computer memory is made out of bits which take the values 0 or 1. This can be achieved by using small capacitors that are either on or off. In articles on the design of dynamic random access memory (DRAM) devices [10], the typical units of capacitance are fF with voltages of the order of 1 V. These capacitors store of the order of 10^{-15} C which represents a few thousands of electrons. Assuming a resistance of the order of 1 Ω, such capacitors can be charged or discharged in about 10^{-15} s which seems compatible with petaFlops operation. It is interesting to consider the limit of miniaturization where electrons, atoms or photons could be manipulated individually [2].

In the following, we will use the term 'quantum computation' in a broad sense, referring either to the manipulations of atoms, electrons, photons or qubits. In general, the basic blocks are associated with a finite Hilbert space. For instance, qubits provide superpositions of two states $|0\rangle$ and $|1\rangle$, while classical computers use devices like capacitors which are either 'on' or 'off'. The general idea is to prepare quantum states with N qubits which can overlap with the 2^N possible states of the Hilbert space. In order to perform a quantum computation, it is crucial to be able to completely discretize the quantization procedure. For an example of quantum circuit building, see IBM Q https://quantum-computing.ibm.com/support/guides/getting-started-with-circuit-composer.

The discretization of space–time is a well-understood procedure used in *lattice field theory*. In this approach, the fields are only defined at space–time points which are expressed as integer multiples of a basic lattice spacing a. This takes care of some of the infinities mentioned above. However, it breaks the continuous space–time symmetries such as translations and rotations and replaces them with discrete ones. If the lattice spacing is a mathematical artifact, it needs to be small compared to any length scale in the problem. We call the limit where that is the case and the lattice spacing becomes irrelevant, the *continuum limit*. In the context of high-energy physics, we are not aware of any evidence for an elementary physical lattice spacing and consequently this limit needs to be taken. On the other hand, for a solid, such a lattice spacing represents a physical effect and should be kept at its physical value.

Lattice field theory is necessary to describe strongly interacting particles with quantum chromodynamics (QCD) and allows us to explain behaviors that cannot be reached in perturbation theory such as the confinement of quarks and gluons.

A simple example of discretized space can be obtained by considering the height of a string element $q(x)$ with respect to its rest height as a function of the distance x from one boundary. To make things simple, we assume that the string vibrates in a plane so that we don't need extra components. We discretize the position of the string element by replacing the continuous position variable x by integer multiples of a small unit of length a and only consider the height of the string at these values of the position:

$$q_n \equiv q(na). \tag{1.3}$$

We are now using a finite number of degrees of freedom and proceed as in classical mechanics.

This is not the end of the story. If the path-integral approach is used we also need to replace the integration over all the field configurations denoted $\int [\mathcal{D}\varphi]$ by discrete sums. As we will see, this can be introduced naturally by using *character expansions* in many situations of interest [11, 12]. This reformulation fits the discrete needs of quantum computing and is relatively easy to visualize. One important goal of these lecture notes is to make this approach accessible to beginners.

1.5 Symmetries and predictive models

The action S is a central ingredient which defines field theory models. Imposing symmetries help us to reduce the number of terms in the action. In addition of space–time symmetries like Lorentz transformations or translations, there are also *internal symmetries* which mix components of fields at a given space time point. Internal symmetries can be global, or local if we let the transformation vary with space and time. Additional conditions on the dimensionality of the terms that can be introduced in the action can be imposed on the basis of perturbative renormalizability or relevance at low energy.

A remarkable achievement of QFT is the standard model of strong and electroweak interactions. The requirements of a $SU(3) \otimes SU(2) \otimes U(1)$ local symmetry, relativistic covariance and the requirement that the terms of the action are products of fields having dimension 4 or less allows us to restrict the number of input parameters to 18 (neglecting the QCD vacuum angle and the masses and mixing of the neutrinos). There is very little room to 'tweak' the model each time a new experiment is completed. This input was used successfully to predict windows for the masses of the W and Z bosons, top quark and Brout–Englert–Higgs boson and predict or confirm more than a thousand pages of data from the Particle Data Group (http://pdg.lbl.gov). A great achievement of the standard model is the prediction of the anomalous magnetic moment of the muon. The discrepancy between theory and experiment is expressed as one digit integers times 10^{-10} (see http://pdg.lbl.gov/2019/reviews/rpp2018-rev-g-2-muon-anom-mag-moment.pdf for a review).

References

[1] Feynman R P 1982 *Int. J. Theor. Phys.* **21** 467–88

[2] Lloyd S 1996 *Science* **273** 1073–8

[3] Bloch I, Dalibard J and Zwerger W 2008 *Rev. Mod. Phys.* **80** 885–964

[4] Debnath S, Linke N M, Figgatt C, Landsman K A, Wright K and Monroe C 2016 *Nature* **536** 63–6

[5] Klco N, Dumitrescu E F, McCaskey A J, Morris T D, Pooser R C, Sanz M, Solano E, Lougovski P and Savage M J 2018 *Phys. Rev.* A **98** 032331

[6] Lamm H and Lawrence S 2018 *Phys. Rev. Lett.* **121** 170501

[7] Gustafson E, Meurice Y and Unmuth-Yockey J 2019 *Phys. Rev.* D **99** 094503

[8] Peskin M E and Schroeder D V 1995 *An Introduction to Quantum Field Theory* (Reading, MA: Addison-Wesley)

[9] Jackson J D 1998 *Classical Electrodynamics* (New York: Wiley)

[10] Luk W, Cai J, Dennard R, Immediato M and Kosonocky S 2006 A 3-transistor dram cell with gated diode for enhanced speed and retention time *2006 Symp. on VLSI Circuits, 2006. Digest of Technical Papers* pp 184–5

[11] Balian R, Drouffe J M and Itzykson C 1975 *Phys. Rev.* D **11** 2104–19

[12] Liu Y, Meurice Y, Qin M P, Unmuth-Yockey J, Xiang T, Xie Z Y, Yu J F and Zou H 2013 *Phys. Rev.* D **88** 056005

IOP Publishing

Quantum Field Theory
A quantum computation approach
Yannick Meurice

Chapter 2

Classical field theory

2.1 Classical action, equations of motion and symmetries

As we have just discussed, the predictive powers of the standard model can be traced to our ability to restrict the number of terms of the action.

The principle of least action appears prominently in the Landau and Lifshitz *Course of Theoretical Physics* [1, 2]. In section 12 (Path integral formulation of field theory) of an influential review article on gauge theories, Abers and Lee [3] provide the quote 'Physics—Where the Action Is' (Anonymous). In classical physics, finding the extrema of the action provides equations of motion which define unambiguously the time evolution of the system. In addition, the invariance of the action under continuous symmetries allows us to identify conserved quantities which, by definition, do not evolve in time. This is guaranteed by Noether's theorem. As we will proceed to explain, equations of motion and conserved quantities are closely related.

In classical mechanics with N coordinates q_j, with $j = 1, \ldots, N$, we will consider the action

$$S = \int_{t_i}^{t_f} dt L(q_j, \dot{q}_j). \tag{2.1}$$

We assume that there is no explicit time dependence in S. Standard examples of explicit time dependence are masses or oscillator frequencies changing with time. If we vary the action

$$\delta S = \int_{t_i}^{t_f} dt \left(\frac{\partial L}{\partial q_j} \delta q_j + \frac{\partial L}{\partial \dot{q}_j} \delta \dot{q}_j \right), \tag{2.2}$$

with implicit sums over the repeated index j and rewrite the integrand as

$$\left(\frac{\partial L}{\partial q_j} - \frac{d}{dt} \frac{\partial L}{\partial \dot{q}_j} \right) \delta q_j + \frac{d}{dt} \left(\frac{\partial L}{\partial \dot{q}_j} \delta q_j \right), \tag{2.3}$$

we can omit the last term which is a total derivative if we assume $\delta q_j = 0$ at t_i and t_f and require $\delta S = 0$, we obtain N Euler–Lagrange equations:

$$\frac{\partial L}{\partial q_j} - \frac{d}{dt}\frac{\partial L}{\partial \dot{q}_j} = 0. \tag{2.4}$$

We now consider special variations that correspond to symmetries of the problem. Simple examples can be found for instance when the q_j are Cartesian coordinates and the problem is invariant under rotations or translations. The simplest situation is when the δq_j leave L and consequently S invariant. It is important to notice that for symmetries $\delta q_j \neq 0$ at the boundary (t_i and t_f). If we perform the symmetry variations as above and apply the equations of motion, then the only parts that remain in the variation of S are the boundary terms:

$$\delta S = 0 = \int_{t_i}^{t_f} dt \frac{d}{dt}\left(\frac{\partial L}{\partial \dot{q}_j}\delta q_j\right) = \frac{\partial L}{\partial \dot{q}_j}\delta q_j \bigg|_{t_i}^{t_f}. \tag{2.5}$$

The invariance guarantees that the two boundary terms at t_i and t_f are the same. This defines a constant of motion. This the basic idea of Noether's theorem. The original article [4] applies to more general situations. The constant of motion is proportional to $\frac{\partial L}{\partial \dot{q}_j}\delta q_j$. We say proportional because this expression contains infinitesimal elements which do not belong to the definition of the charge. This dependence can be removed by taking derivatives with respect to the infinitesimal elements.

For a Lagrangian not depending explicitly on time, we can shift the time variable and identify a constant of motion which turns out to be the energy. Under a shift in time $t \to t + \epsilon$, the action 'loses' a small contribution near t_i and 'receives' a small contribution near t_f:

$$\delta S = \epsilon L|_{t_i}^{t_f} = \epsilon \int_{t_i}^{t_f} dt \frac{d}{dt} L. \tag{2.6}$$

On the other hand, we can also calculate δS as before using $\delta q_j = \epsilon \dot{q}_j$. After applying the equations of motion and comparing the two expressions of δS we find that the Hamiltonian

$$H = \frac{\partial L}{\partial \dot{q}_j}\dot{q}_j - L, \tag{2.7}$$

is the same at t_i and t_f. Notice that in equation (2.6) the Lagrangian changes by a total derivative and the action by boundary terms, in contrast to the first example where there was a strict invariance.

The Hamiltonian formalism introduces momenta $p_j = \frac{\partial L}{\partial \dot{q}_j}$ and converts the N Euler–Lagrange equations into $2N$ first-order equations which are easier to understand from the point of view of ordinary differential equations. It plays a significant role in the quantization program. Students should be fluent with Hamilton equations, Poisson brackets and canonical transformations as discussed for instance in reference [1].

Example: Consider the case $L = (1/2)(\dot{q}_1^2 + \dot{q}_2^2)$. L is invariant under the transformation $\delta q_1 = \epsilon q_2$ and $\delta q_2 = -\epsilon q_1$ and $((\partial L/\partial \dot{q}_j)\delta q_j) = \epsilon(\dot{q}_1 q_2 - \dot{q}_2 q_1)$. We can define $A = \dot{q}_1 q_2 - \dot{q}_2 q_1$ which should be a conserved quantity. If we use the equation of motion $\ddot{q}_j = 0$, $\dot{A} = \dot{q}_1 \dot{q}_2 - \dot{q}_2 \dot{q}_1 = 0$. If q_1 and q_2 are interpreted as Cartesian coordinates in a plane, A is the angular momentum in the normal direction. The coordinates q_j do not appear explicitly in L and are called ignorable. From the equations of motion we see that $\frac{\partial L}{\partial q_j} = 0$ implies that the momentum $\frac{\partial L}{\partial \dot{q}_j} = \dot{q}_j$ is constant. This can also be obtained by the Noether method by noticing the invariance of L under shifts of the q_j. In addition, there is no explicit dependence on time and the application of equation (2.7) provides the conservation of energy which in this case is purely kinetic.

Notice that in this example the energy conservation follows from the momentum conservation. In general, how many independent constants of motion can we identify? From the theory of ordinary differential equations [5], we know that we can locally 'rectify' the $2N$ Hamilton equations. One coordinate plays a role similar to time while the $2N - 1$ other coordinates are kept constant. So it seems that $2N - 1$ should be an upper bound on the number of constants of motion. This bound is saturated in the previous example. In the quantum version, 'independent' should be replaced by 'commuting'. With this requirement, if we can use quantum versions of action-angle variable, there should be at most N commuting operators. In this limiting situation, the system is called integrable. In classical mechanics, we can replace 'commuting' by 'having vanishing Poisson brackets'. Classical integrable systems were studied by Liouville [5].

Exercise 1: Consider N harmonic oscillators with identical frequencies ω. The Lagrangian reads

$$L = \sum_{j=1}^{N} \left(\frac{1}{2}\dot{q}_j^2 - \frac{1}{2}\omega^2 q_j^2 \right). \tag{2.8}$$

Calculate the equations of motion, enumerate the continuous symmetries and their associated conserved quantities constructed using Noether's theorem. Using the equations of motion, check their conservation.

Solution. The N Euler–Lagrange equations are:

$$\frac{\partial L}{\partial q_j} - \frac{d}{dt}\frac{\partial L}{\partial \dot{q}_j} = -\ddot{q}_j - \omega^2 q_j = 0.$$

The Lagrangian is invariant under $O(N)$ symmetries (see appendix A) which are 'rotating' the q_j forming a N-dimensional vector. $O(N)$ has $\frac{N(N-1)}{2}$ infinitesimal generators which for any pair of distinct indices i and j provide the transformation:

$$\delta q_i = \epsilon q_j,$$
$$\delta q_j = -\epsilon q_i.$$

The corresponding Noether charges are

$$Q_{ij} = \dot{q}_j q_i - \dot{q}_i q_j. \tag{2.9}$$

These are constants of motion because

$$\dot{Q}_{ij} = \ddot{q}_j q_i - \ddot{q}_i q_j = -\omega^2(q_j q_i - q_i q_j) = 0. \tag{2.10}$$

The equations of motion have been used to obtain the second equality. In addition, time does not appear explicitly in L, so the energy is conserved:

$$H = \frac{\partial L}{\partial \dot{q}_j} \dot{q}_j - L = \sum_{j=1}^{N}\left(\frac{1}{2}\dot{q}_j^2 + \frac{1}{2}\omega^2 q_j^2\right), \tag{2.11}$$

is a constant of motion because

$$\dot{H} = \sum_{j=1}^{N}(\dot{q}_j \ddot{q}_j + \omega^2 q_j \dot{q}_j) = 0, \tag{2.12}$$

after using the equation of motion for the first term.

2.2 Transition to field theory

The transition to field theory can be accomplished by replacing the discrete index j by a spatial index $\mathbf{x}$:

$$j \to \mathbf{x}$$
$$\sum_j \to \int d^{D-1}x \tag{2.13}$$
$$q_j(t) \to \phi(\mathbf{x}, t)$$
$$\dot{q}_j(t) \to \partial_\mu \phi(\mathbf{x}, t).$$

In field theory, the action S is the integral over the D-dimensional space–time of a Lagrangian density $\mathcal{L}(x)$ which we write as a function of the fields $\varphi(x)$ (this replaces $q_j(t)$) or their D-gradient $\partial_\mu \varphi(x) \equiv \frac{\partial}{\partial x^\mu}\varphi(x)$ (this replaces $\dot{q}_j(t)$). In section 2.3 we will explain why we use this particular form.

It is relatively easy to describe the evolution of a point particle in time. We can follow the trajectories in space and infer the velocities using the tangent vectors. Assuming that second order Newton equations hold, the evolution is completely fixed by the initial position and velocity while higher derivatives are obtained from the equations of motion. It is thus easier to use first order equations in a phase space which has twice the dimensions of the original space as done in the Hamiltonian formalism. It is more difficult to visualize the 'classical trajectories' of a field. They interpolate between the field configurations on the initial and final 'time slices' $\phi(\mathbf{x}, t_i)$ and $\phi(\mathbf{x}, t_f)$. If we assume that these functions of $\mathbf{x}$ are smooth, slowly varying and have periodic boundary conditions, we could try to approximate them using Fourier series. A more generic method consists in averaging the fields in (hyper)cubes of spatial volume a^{D-1}, which brings us back to the q_j formulation with a (hyper)cubic lattice structure. A simple example in $D = 2$, so one spatial dimension is illustrated in equation (1.3).

Applying a variation $\delta\varphi(x)$ to the action

$$S = \int d^D x \mathcal{L}(x), \tag{2.14}$$

and proceeding as in classical mechanics with the replacements mentioned above, we get the equations of motion

$$\partial \mathcal{L}/\partial \varphi(x) - \partial_\mu(\partial \mathcal{L}/\partial(\partial_\mu \varphi)) = 0. \tag{2.15}$$

Similarly, we find that for a field transformation corresponding to an internal symmetry leaving $\mathcal{L}$ invariant, we have a Noether current $J^\mu \propto (\partial \mathcal{L}/\partial(\partial_\mu \varphi))\delta\varphi$ (with possible sums over various indices not written explicitly) which is conserved ($\partial_\mu J^\mu = 0$). For space–time symmetries, $\mathcal{L}$ changes by a D-divergence and there is an extra term as for the Hamiltonian in classical mechanics. Examples will be given in section 2.4.

2.3 Symmetries

The symmetries of the action either refer to transformations of space and time (special relativity and translations) or transformations of internal degrees of freedom (charge, isospin, color, ...). In addition, a symmetry is called global when the transformations are identical at every space–time point and local otherwise.

Relativistic covariance is one of the basic principles in many branches of theoretical physics. It asserts that the laws of physics and the speed of light should be the same in any inertial frame. Maxwell's equations are consistent with this general requirement and played an important role in the development of special relativity. The fact that the speed of light c has an absolute meaning makes it convenient to pick a system of units where $c = 1$. The current definition of the meter is the distance covered by light in $1/299\,792\,458$ s.

In the context of high-energy physics (HEP), $D = 4$ should be understood as 3 space and one time dimensions. The special relativity transformations (three boosts and three rotations) are well-known, however, it is straightforward to formulate special relativity for any $D \geqslant 2$.

In the rest of this chapter, we follow PS notations. A D-vector x^μ with the index up is a short notation for $(ct, x_1, \dots, x_{D-1})$. Soon we will drop the c. The $D - 1$-dimensional spatial vectors $(x_1, \dots, x_{D-1})$ will be denoted $\mathbf{x}$. We call x^μ a *contravariant vector*. It transforms like

$$x^\mu \to x'^\mu = \Lambda^\mu_\nu x^\nu, \tag{2.16}$$

under Lorentz transformations. These transformations form a group that will be defined below in equation (2.27).

A scalar field is a field which transforms in a trivial way:

$$\phi(x) \rightarrow \phi'(x') = \phi(x). \qquad (2.17)$$

It has a single component. Later we will define multicomponent fields which transform into linear combinations of the components and will be called vector, spinor or tensor fields. The general form of the transformation is then:

$$\phi^A(x) \rightarrow \phi'^A(x') = M^A_B(\Lambda)\phi^B(x), \qquad (2.18)$$

with the A and B corresponding to one of the possibility listed above (e.g. a spinor index). Examples of $M^A_B(\Lambda)$ will be given in the following sections. As we will see, the Lorentz transformations form a group and we will require that the multiplication of the matrices M is consistent with the group composition, namely

$$M^A_B(\Lambda)M^B_C(\Lambda') = M^A_C(\Lambda\Lambda'). \qquad (2.19)$$

When this property is satisfied, we say that M is a *representation* of the group. Finding all the possible representations of the Lorentz group is a problem that is completely solved [6].

We now consider the D-gradient of a scalar field $\frac{\partial}{\partial x^\mu}\phi(x)$. Under a Lorentz transformation:

$$\frac{\partial}{\partial x^\mu}\phi(x) \rightarrow \frac{\partial}{\partial x'^\mu}\phi'(x') = \frac{\partial}{\partial x'^\mu}\phi(x) = \frac{\partial x^\nu}{\partial x'^\mu}\frac{\partial}{\partial x^\nu}\phi(x), \qquad (2.20)$$

with summation over the repeated index ν kept implicit. Since

$$\frac{\partial x'^\nu}{\partial x^\mu} = \Lambda^\nu_\mu, \qquad (2.21)$$

and the application of the chain rule yields

$$\frac{\partial x'^\nu}{\partial x^\mu}\frac{\partial x^\mu}{\partial x'^\rho} = \delta^\nu_\rho, \qquad (2.22)$$

we see that $\frac{\partial}{\partial x^\mu}\phi(x)$ transforms with the inverse of Λ^ν_μ. We call any D-vector y_μ transforming like a D-gradient a *covariant vector*. By contracting (summing over the indices) a contravariant and a covariant vector, we form a Lorentz-invariant quantity also called a Lorentz scalar. A simple example is $\frac{\partial}{\partial x^\mu}x^\mu = D$. In the following, we use the notation ∂_μ for $\partial/\partial x^\mu$.

We now introduce the Lorentz metric tensor $g_{\mu\nu}$ which converts a contravariant vector into a covariant one:

$$x_\mu = g_{\mu\nu}x^\nu. \qquad (2.23)$$

We can construct a Lorentz scalar by contracting two contravariant vectors with the metric: $y^\mu g_{\mu\nu} x^\nu$. In addition, we can also raise the indices using the contravariant form of the metric $g^{\mu\nu}$ such that $g^{\mu\nu} g_{\nu\rho} = \delta^\mu_\rho$.

Light-like events are characterized by

$$c^2 dt^2 - d\mathbf{x} \cdot d\mathbf{x} = 0, \tag{2.24}$$

a property that should be preserved by Lorentz transformations according to the basic principle of relativistic covariance. Consequently, we can write this invariant quantity as

$$dx^\mu g_{\mu\nu} dx^\nu = c^2 dt^2 - d\mathbf{x} \cdot d\mathbf{x}, \tag{2.25}$$

provided that $g_{\mu\nu}$ is diagonal with entries $(1, -1, -1, \ldots -1)$. For instance for $D = 4$, the metric can be visualized as the matrix elements of

$$G = \begin{pmatrix} 1 & 0 & 0 & 0 \\ 0 & -1 & 0 & 0 \\ 0 & 0 & -1 & 0 \\ 0 & 0 & 0 & -1 \end{pmatrix}. \tag{2.26}$$

We are now in position to define the Lorentz group as the set of transformations that leave the Lorentz metric invariant:

$$\Lambda^{\mu'}_\mu g_{\mu'\nu'} \Lambda^{\nu'}_\nu = g_{\mu\nu}. \tag{2.27}$$

Exercise 2: Show that these transformations form a group. Note: the concept of group is defined in appendix A. Hint: write equation (2.27) as a matrix equation. Enumerate the infinitesimal generators of the group and their algebra. Discuss their physical meaning.

Solution. For convenience, we can think of $x'^\mu = \Lambda^\mu_\nu x^\nu$ as a $D \times D$ matrix Λ acting on a D-dimensional column vector. The upper index μ in Λ is then interpreted as a row index and ν as a column index. Similarly, we treat $g_{\mu\nu}$ as a $D \times D$ diagonal matrix G with entries $(1, -1, -1, \ldots -1)$. With these notations equation (2.27) becomes

$$\Lambda^T G \Lambda = G. \tag{2.28}$$

We can prove that all the matrices Λ satisfying this property form a group: (i) if $\Lambda_i^T G \Lambda_i = G$ for $i = 1, 2$, then $\Lambda_1 \Lambda_2$ also satisfies the condition (2.28):

$$(\Lambda_1 \Lambda_2)^T G (\Lambda_1 \Lambda_2) = \Lambda_2^T \Lambda_1^T G \Lambda_1 \Lambda_2 = \Lambda_2^T G \Lambda_2 = G; \tag{2.29}$$

(ii) the matrix multiplication is associative; (iii) the identity matrix clearly satisfies the condition (2.28); (iv) if Λ satisfies equation (2.28), so does its inverse. In order to prove this, we use the fact that $G = G^{-1} = G^T$ which, used with equation (2.28), implies that

$$\Lambda^{-1} = G \Lambda^T G, \tag{2.30}$$

and

$$(\Lambda^{-1})^T G \Lambda^{-1} = (G\Lambda G)G(G\Lambda^T G) = G\Lambda(G\Lambda^T G) = G\Lambda\Lambda^{-1} = G. \tag{2.31}$$

The infinitesimal generators of the Lorentz group can be obtained from the first order expansion:

$$\Lambda^{\mu}_{\nu} \simeq \delta^{\mu}_{\nu} + \epsilon^{\mu}_{\nu}. \tag{2.32}$$

After plugging into equation (2.28) and expanding to first order, we obtain

$$\epsilon_{\mu\nu} + \epsilon_{\nu\mu} = 0, \tag{2.33}$$

with $\epsilon_{\mu\nu} \equiv g_{\mu\rho}\epsilon^{\rho}_{\nu}$ which must be antisymmetric in its two indices. The $D - 1$ parameters ϵ_{0i} correspond to boosts in the ith spatial direction. The ϵ_{ij} with i and j spatial indices, correspond to rotations in the $\frac{(D-1)(D-2)}{2}$ independent planes. Note that it is common to use Greek symbols for space–time indices and Latin indices for space indices. However, not all Latin indices should be interpreted as spatial indices (they can also be used for internal degrees of freedom).

We now specialize the discussion to $D = 4$. There are three boosts and three rotations. With the matrix convention developed for the proof of the group properties, an infinitesimal boost in the z direction requires $\epsilon^0_3 = \epsilon^3_0 = \epsilon$. Lowering with the metric, we get $\epsilon_{03} = -\epsilon_{30} = \epsilon$ consistently with the antisymmetric condition. The infinitesimal form can be exponentiated for a general hyperbolic angle η, also called rapidity, with result

$$\Lambda_{z-\text{boost}} = \begin{pmatrix} \cosh\eta & 0 & 0 & \sinh\eta \\ 0 & 1 & 0 & 0 \\ 0 & 0 & 1 & 0 \\ \sinh\eta & 0 & 0 & \cosh\eta \end{pmatrix}. \tag{2.34}$$

If we apply this transformation to a particle of mass m at rest ($p^0 = m$, $p^i = 0$ in $c = 1$ units), we obtain $E = m\cosh\eta$ and $p^3 = m\sinh\eta$. This is consistent with the conventional definition of rapidity http://pdg.lbl.gov/2019/reviews/rpp2018-rev-kinematics.pdf.

Similarly, we obtain infinitesimal rotations about the z direction with $\epsilon^2_1 = -\epsilon^1_2 = \epsilon$. Lowering with the metric, we get $\epsilon_{12} = -\epsilon_{21} = \epsilon$ consistently with the antisymmetric condition. The infinitesimal form can be exponentiated for a general rotation angle θ, with result

$$\Lambda_{z-\text{rotation}} = \begin{pmatrix} 1 & 0 & 0 & 0 \\ 0 & \cos\theta & -\sin\theta & 0 \\ 0 & \sin\theta & \cos\theta & 0 \\ 0 & 0 & 0 & 1 \end{pmatrix}. \tag{2.35}$$

We just discussed the transformation properties of contravariant and covariant vectors. Similarly, we can define *tensors* which are objects with multiple contravariant and covariant indices each transforming in the same way as the above-mentioned vectors. A simple example is $T^{\mu\nu}_\rho$, a tensor with two contravariant and one covariant indices. It transforms like

$$T^{\mu\nu}_\rho \rightarrow T^{\mu'\nu'}_{\rho'} \frac{\partial x'^\mu}{\partial x^{\mu'}} \frac{\partial x'^\nu}{\partial x^{\nu'}} \frac{\partial x^{\rho'}}{\partial x'^\rho}. \tag{2.36}$$

Be careful about primes. In this equation, the primed indices are dummy indices. The prime on x indicates if the transformation is for a contravariant index (x' on top) or covariant (x' at the bottom).

In contrast to space–time symmetries, internal symmetries only affect the various components of a field at a given space–time point. The generic form for a global internal symmetry is

$$\phi^i(x) \rightarrow \phi'^i(x) = U^i_j \phi^j(x), \tag{2.37}$$

with i and j internal indices such as the isospin and U unitary or orthogonal matrices. The global symmetry can be promoted to a local symmetry if we allow U^i_j to become a function of the space–time point x: $U^i_j(x)$. One well known example is when i and j correspond to the color indices of QCD.

2.4 The Klein–Gordon field

The simplest relativistic field theory that we can build involves a single real scalar field $\phi(x)$. A Lorentz-invariant Lagrangian density can be built out of arbitrary powers of ϕ and $\partial_\mu \phi \partial^\mu \phi$. In addition, it is common to require the invariance under $\phi \rightarrow -\phi$ which excludes odd powers of ϕ. The simplest case is the quadratic (only two powers of the fields) density

$$\mathcal{L}^{KG} = \frac{1}{2}\partial_\mu \phi \partial^\mu \phi - \frac{1}{2}(m^2 c^2/\hbar^2)\phi^2. \tag{2.38}$$

In order to apply equation (2.15), we notice that due to the symmetry under the exchange of the two indices of $g^{\mu\nu}$,

$$\delta(\partial_\mu \phi \partial^\mu \phi) = \delta(\partial_\mu \phi g^{\mu\nu} \partial_\nu \phi) = 2\delta(\partial_\mu \phi)g^{\mu\nu} \partial_\nu \phi = 2\delta(\partial_\mu \phi)\partial^\mu \phi, \tag{2.39}$$

and obtain the equation of motion

$$\partial_\mu \partial^\mu \phi + (m^2 c^2/\hbar^2)\phi = 0, \tag{2.40}$$

which is called the Klein–Gordon (KG) equation. This is a linear equation and we call the corresponding solutions 'free fields'. The interactions are introduced by nonlinear terms. The differential operator $\partial_\mu \partial^\mu$ is called the d'Alembertian and denoted $\Box$. It appears in wave equations and its explicit form in space–time coordinates is

$$\Box = \partial^2/c^2 \partial t^2 - \nabla \cdot \nabla. \tag{2.41}$$

An interactive theory can be defined by adding higher order terms to $\mathcal{L}$ such as $-\lambda\phi^4$. This leads to nonlinear equations of motion. This type of field theory is often called a 'lambda phi four' theory.

The KG equation admits complex solutions of the form

$$\phi^{\pm}(x) \equiv u e^{\mp i x \cdot p / \hbar}, \tag{2.42}$$

with $x \cdot p \equiv x_\mu p^\mu$. Plugging into equation (2.40), we get the condition

$$p^2 \equiv p_\mu p^\mu = m^2 c^2, \tag{2.43}$$

which we call the relativistic mass-shell condition. p^μ is a short notation for the D-vector $(E/c, \mathbf{p})$ with E the energy and $\mathbf{p}$ the spatial momenta.

At this point we realize that the powers of c and $\hbar$ appearing in the equations can be trivially figured out by dimensional analysis. In the following we will use the 'natural units' where $c = 1$ and $\hbar = 1$. In these units one second and 299 792 458 m represent the same physical situation. The standard conversion to energy is

$$\hbar c = 197.3269788(12) \text{ MeV fm}, \tag{2.44}$$

with 1 fm $= 10^{-15}$ m. For more conversions see http://pdg.lbl.gov/2019/reviews/rpp2018-rev-phys-constants.pdf.

The KG action is not only invariant under Lorentz transformations, it is also invariant under space–time translations. For $D = 4$, this leads to 6 Noether currents for the Lorentz transformations and 4 for the translations (one corresponds to the energy and three to the momenta).

Exercise 3: Work out the Noether current and the conserved quantities corresponding to the space–time translations for KG in arbitrary dimensions D.

Solution. See PS p 19.

It is easy to generalize the KG action to complex fields. Writing the complex field $\Phi = (\phi_1 + i\phi_2)/\sqrt{2}$ (the $\sqrt{2}$ is an historical annoyance) in terms of two real fields ϕ_j:

$$\mathcal{L}^{CKG} = \partial_\mu \Phi \partial^\mu \Phi^\star - m^2 \Phi \Phi^\star = \frac{1}{2}\partial_\mu \phi_j \partial^\mu \phi_j - \frac{1}{2}m^2 \phi_j \phi_j, \tag{2.45}$$

with a sum over $j = 1, 2$ implicit. $\mathcal{L}^{CKG}$ is just the sum of two copies of $\mathcal{L}^{KG}$ with real fields. This model has an internal $O(2)$ symmetry which rotates the two fields as the components of a two-dimensional vector. A rotation by an angle θ can also be expressed as a multiplication of the complex field Φ by a phase $e^{i\theta}$. In both cases we find

$$\phi_1 \rightarrow \cos(\theta)\phi_1 - \sin(\theta)\phi_2, \text{ and } \phi_2 \rightarrow \cos(\theta)\phi_2 + \sin(\theta)\phi_1. \tag{2.46}$$

If we replace the sum over $j = 1, 2$ by a sum over N real components $j = 1, 2, \ldots, N$, then the model has a $O(N)$ symmetry. $O(N)$ is the group of orthogonal matrices of dimension N. These matrices are real N by N matrices A such that $AA^T = \mathbb{1}$.

Exercise 4: Show that the orthogonal N by N matrices form a group. Enumerate the infinitesimal generators and show that they close under commutation.

Solution. See appendix A.

Interesting theories with a $O(N)$ symmetry can be obtained by adding powers of $\phi_j\phi_j$. For instance,

$$\mathcal{L}^{O(N)} = \frac{1}{2}\partial^\mu\phi_j\partial_\mu\phi_j - \lambda(\phi_j\phi_j - v^2)^2, \qquad (2.47)$$

has interesting properties that will be discussed later.

2.5 The Dirac field

The Dirac field satisfies the Dirac equation

$$(i\partial_\mu\gamma^\mu - m)\psi = 0. \qquad (2.48)$$

ψ is often called a 'Dirac spinor'. The γ^μ are dimension-dependent matrices which satisfy the anticommutation relations

$$\{\gamma^\mu, \gamma^\nu\} = \gamma^\mu\gamma^\nu + \gamma^\nu\gamma^\mu = 2g^{\mu\nu}\mathbb{1}. \qquad (2.49)$$

As the two gradients $\partial_\mu\partial_\nu$ commute and we can change the names of 'dummy' indices ($y_\mu x^\mu = y_\nu x^\nu$) and obtain

$$\partial_\mu\gamma^\mu\partial_\nu\gamma^\mu = \frac{1}{2}\partial_\mu\partial_\nu\{\gamma^\mu, \gamma^\nu\} = \Box\mathbb{1}. \qquad (2.50)$$

The operator $\partial_\mu\gamma^\mu$ is denoted $\partial\!\!\!/$ and is in some sense the 'square root of the d'Alembertian' (but it is a diagonal matrix of differential operators). The main result of this section is that given any Lorentz transformation Λ^ν_μ, it is possible to construct matrices $\Lambda_{1/2}$ such that

$$\Lambda_{1/2}^{-1}\gamma^\mu\Lambda_{1/2} = \Lambda^\mu_\nu\gamma^\nu. \qquad (2.51)$$

Exercise 5: Show that if we define

$$\psi'(x') = \Lambda_{1/2}\psi(x), \qquad (2.52)$$

equation (2.51) implies that the Dirac equation has the same form in any inertial frame.

Solution. See PS p 42.

The γ^μ form what we call a Clifford Algebra defined by equation (2.49). The algebra can be realized with matrices of dimensions $2^{D/2}$ for D even and $2^{(D-1)/2}$ for D odd. For $D = 2$ and 3 we can use the Pauli matrices since they anticommute and multiply by i for the space-like indices in order to get the minus signs of the metric. For instance, for $D = 2$, we can choose $\gamma^0 = \sigma^1$ and $\gamma^1 = i\sigma^2$. With this choice,

$$\not{\partial}_{D=2} = \begin{pmatrix} 0 & \partial_0 + \partial_1 \\ \partial_0 - \partial_1 & \end{pmatrix}. \tag{2.53}$$

For $D = 3$, we just add $\gamma^2 = i\sigma^3$. Notice that $\gamma^2 = -i\gamma^0\gamma^1$. For $D = 4$, it is common to use the so-called chiral basis (also called the Weyl basis) where, following the PS convention, we have:

$$\gamma^\mu = \left(\begin{array}{c|c} 0 & \sigma^\mu \\ \hline \bar{\sigma}^\mu & 0 \end{array} \right), \tag{2.54}$$

where σ^μ is a four-vector with matrix components $(\mathbb{1}_2, \boldsymbol{\sigma})$ while $\bar{\sigma}^\mu$ has matrix components $(\mathbb{1}_2, -\boldsymbol{\sigma})$. These γ^μ can be written as *direct products* of Pauli matrices:

$$\gamma^0 = \sigma^1 \otimes \mathbb{1}_2 \text{ and } \gamma^j = i\sigma^2 \otimes \sigma^j. \tag{2.55}$$

This means that we replace the ones in the matrix on the left of $\otimes$ by the matrix on the right. The matrices are then multiplied component by component:

$$(A \otimes B)(C \otimes D) = AC \otimes BD. \tag{2.56}$$

By putting together several Hilbert spaces we obtain a new Hilbert space often called the 'tensor product':

$$|i_1, i_2, \dots i_M\rangle = |i_1\rangle \otimes |i_2\rangle \dots |i_m\rangle. \tag{2.57}$$

The matrix elements are then

$$\langle j_1, j_2, \dots j_M | A_1 \otimes A_2 \otimes \cdots A_M | i_1, i_2, \dots i_M \rangle = \langle j_1 | A_1 | i_1 \rangle \langle j_2 | A_2 | i_2 \rangle \dots \langle j_M | A_M | i_M \rangle. \tag{2.58}$$

Exercise 6: Use the component-by-component multiplication rule to check that the matrices (2.55) satisfy the Clifford algebra relations in equation (2.49).

In $D = 2$, it is possible to construct one extra matrix that anticommutes with all the existing ones by taking the product of these matrices. A similar construction is possible in $D = 4$ and we call this fifth matrix γ^5. We follow the phase convention of PS and define

$$\gamma^5 \equiv i\gamma^0\gamma^1\gamma^2\gamma^3. \tag{2.59}$$

It is clear that it anticommutes with the four γ^μ because as we move any γ^μ from the left to the right of γ^5, it will commute with itself and anticommute with the three others. The construction extends to any even dimension. With the choice given in equation (2.54), γ^5 is diagonal and reads

$$\gamma^5 = \left(\begin{array}{c|c} -\mathbb{1}_2 & 0 \\ \hline 0 & +\mathbb{1}_2 \end{array} \right). \tag{2.60}$$

It is clear that $(\gamma^5)^2 = \mathbb{1}_4$. We can now define 'Left' and 'Right' projectors

$$P_L = (\mathbb{1} - \gamma^5)/2 \text{ and } P_R = (\mathbb{1} + \gamma^5)/2, \tag{2.61}$$

which satisfy the relations $P_L^2 = P_L$, $P_R^2 = P_R$ and $P_L P_R = 0$. Looking at the explicit form of γ^5, we can separate the four-dimensional Dirac spinor into two (a Left and a Right) two-dimensional spinors (called Weyl spinors):

$$\psi = \begin{pmatrix} \psi_L \\ \psi_R \end{pmatrix}, \tag{2.62}$$

or equivalently

$$P_L\psi = \begin{pmatrix} \psi_L \\ 0 \end{pmatrix} \text{ and } P_R\psi = \begin{pmatrix} 0 \\ \psi_R \end{pmatrix}. \tag{2.63}$$

These spinors are often called 'left-handed' or 'right-handed' while γ^5 generates 'chiral' transformation (from the Greek $\chi\epsilon\iota\rho$ ('kheir') meaning hand) because the projection of the spin (not yet defined) of massless fermions on their momentum (called helicity) is sometimes remembered using left- or right-hand rules. The concept of helicity is discussed in PS (p 47).

We now have the tools to construct the $\Lambda_{1/2}$ satisfying the Lorentz-covariance condition (2.51). The discussion will focus on $D = 4$. An arbitrary Lorentz transformation acting on a four-vector can be written as

$$\Lambda_\beta^\alpha = \left(\exp\left(-\frac{i}{2}\omega_{\mu\nu}\mathcal{J}^{\mu\nu} \right) \right)_\beta^\alpha, \tag{2.64}$$

with

$$(\mathcal{J}^{\mu\nu})_\beta^\alpha = i\left(g^{\mu\alpha}\delta_\beta^\nu - g^{\nu\alpha}\delta_\beta^\mu \right), \tag{2.65}$$

α and β are matrix indices. With this parametrization, we have

$$\Lambda_{1/2} = \exp((1/8)\omega_{\mu\nu}[\gamma^\mu, \gamma^\nu]). \tag{2.66}$$

Exercise 7: Work out the details of this solution for the special cases of a rotation about the z-axis and a Lorentz boost along the z-axis. Start with the infinitesimal form and extend the results by matrix exponentiation. Useful identity: if B and C are arbitrary matrices: $\exp(-B)C\exp(B) = C + [C, B] + \frac{1}{2!}[[C, B], B] + \frac{1}{3!}[[[C, B], B], B] + \cdots$.

We have seen in *exercise* 5 that under a Lorentz transformation

$$(i\partial_\mu\gamma^\mu - m)\psi \to \Lambda_{1/2}(i\partial_\mu\gamma^\mu - m)\psi. \tag{2.67}$$

In order to construct an invariant action, we need to invent an object which transforms with $(\Lambda_{1/2})^{-1}$. If $\Lambda_{1/2}$ was an unitary transformation, we could just take $\psi^\dagger$, however this is not the case for arbitrary transformations. This can be traced to the fact that γ^0 is Hermitian while the γ^i, $i = 1, 2, 3$ are anti-Hermitian. This can be summarized as

$$\gamma^{\mu\dagger} = \gamma^0\gamma^\mu\gamma^0. \tag{2.68}$$

Using this relation it is possible to show that

$$\gamma^0 (\Lambda_{1/2})^\dagger \gamma^0 = (\Lambda_{1/2})^{-1}. \tag{2.69}$$

Exercise 8: Prove this relation. Useful identity: for an invertible matrix A and an arbitrary matrix B: $A \exp(B) A^{-1} = \exp(ABA^{-1})$.

Using this result, we define

$$\bar{\psi} \equiv \psi^\dagger \gamma^0, \tag{2.70}$$

and find that under a Lorentz transformation

$$\bar{\psi} \to \bar{\psi} (\Lambda_{1/2})^{-1}. \tag{2.71}$$

This allows us to write the Lorentz-invariant Lagrangian density

$$\mathcal{L}^{\text{Dirac}} = \bar{\psi} (i \partial_\mu \gamma^\mu - m) \psi. \tag{2.72}$$

It is interesting to rewrite $\mathcal{L}^{\text{Dirac}}$ in term of the Weyl spinors. The result is

$$\mathcal{L}^{\text{Weyl-Dirac}} = \psi_L^\dagger i \bar{\sigma}^\mu \partial_\mu \psi_L + \psi_R^\dagger i \sigma^\mu \partial_\mu \psi_R - m \psi_L^\dagger \psi_R - m \psi_R^\dagger \psi_L. \tag{2.73}$$

The Dirac Lagrangian density $\mathcal{L}^{\text{Dirac}}$ has internal symmetries. First we can multiply ψ by a global phase so that

$$\psi \to e^{i\alpha} \psi \text{ and } \bar{\psi} \to e^{-i\alpha} \bar{\psi}. \tag{2.74}$$

This symmetry is related to the charge conservation and clearly leaves $\mathcal{L}^{\text{Dirac}}$ unchanged. It is often called a 'vector symmetry' because it treats left and right Weyl spinors in a parity invariant way.

We can promote this symmetry to a local symmetry where α becomes a space–time-dependent phase $\alpha(x)$ provided that we replace ∂_μ by the covariant derivative D_μ defined as

$$D_\mu \equiv \partial_\mu + i A_\mu(x), \tag{2.75}$$

with

$$A_\mu(x) \to A_\mu(x) - \partial_\mu \alpha(x). \tag{2.76}$$

With this extra term,

$$D_\mu \psi \to e^{i\alpha(x)} D_\mu \psi, \tag{2.77}$$

and the Lagrangian density is invariant under the local symmetry. Later, we will say that the photon couples to a conserved current and introduce a coupling constant.

When $m = 0$, we have an extra global symmetry called *chiral symmetry*. We can perform a chiral transformation

$$\psi \to e^{+i\beta\gamma^5} \psi \text{ and } \bar{\psi} \to \bar{\psi} e^{+i\beta\gamma^5}, \tag{2.78}$$

which leaves $\mathcal{L}^{\text{Dirac}}$ unchanged. Notice the plus sign in the exponent for the transformation of $\bar{\psi}$. This is due to the γ^0 in the definition (2.70) of $\bar{\psi}$. To prove the invariance, we used the identities

$$e^{-i\beta\gamma^5}\gamma^0 = \gamma^0 e^{+i\beta\gamma^5} \text{ and } e^{+i\beta\gamma^5}\slashed{\partial}e^{+i\beta\gamma^5} = \slashed{\partial}. \tag{2.79}$$

The chiral symmetry is often called an 'axial symmetry' because it gives the left and right Weyl spinors opposite phases. From the explicit form of γ^5 (2.60), we see that the transformation (2.78) translates into

$$\psi_L \to e^{-i\beta}\psi_L \text{ and } \psi_R \to e^{i\beta}\psi_R. \tag{2.80}$$

We can combine the axial and vector transformations to obtain transformations that act only on ψ_L ($\alpha = -\beta$) or only on ψ_R ($\alpha = \beta$). The fact that in the massless limit, independent phase transformations may be applied to ψ_L and ψ_R is manifest from the expression (2.73).

The understanding of Weyl spinors is crucial for the standard model. In the electro-weak sector, left and right spinors interact differently with the W and Z bosons. For QCD, if we only consider the light up and down quarks, their contributions to the mass of the neutron and protons only account for a few percent of the observed mass. The chiral limit where the masses of the up and down quarks is set to zero is a good approximation of the physics because chiral symmetry is broken 'dynamically' by the strong interactions which account for most of the mass of the proton and neutron.

Exercise 9: Consider a model with N_f massless Dirac fermions (we can think of these N_f species as 'flavors'). Write $\mathcal{L}$ in terms of Weyl spinors. What are the internal symmetries of $\mathcal{L}$?

The Dirac equation is a linear equation which is closely related to the KG equation. Multiplying the Dirac equation (2.48) by $(-i\slashed{\partial} - m)$ and using equation (11.32) we obtain the KG equation:

$$(-i\partial_\mu\gamma^\mu - m)(i\partial_\mu\gamma^\mu - m)\psi = (\Box + m^2)\psi = 0. \tag{2.81}$$

The solutions of Dirac equations can be constructed as solutions of the KG equation (2.42) multiplied by a four-dimensional vector depending only on the four-momenta appearing in the KG solution. The conventional notations are

$$u(p)e^{-ip\cdot x}, \tag{2.82}$$

for the solutions with 'positive frequencies' and

$$v(p)e^{+ip\cdot x}, \tag{2.83}$$

for the solutions with 'negative frequencies'. In these expressions, it is understood that the four-momenta satisfy the mass-shell condition $p^2 = m^2$. We will come back on the physical interpretation, related to antiparticles, when we discuss the quantum states associated with these classical solutions. Plugging these expressions in the Dirac equation, we see that the column matrices must satisfy the equations

$$(\not{p} - m)u(p) = 0 \text{ and } (\not{p} + m)v(p) = 0. \tag{2.84}$$

Explicit forms of the solutions can be found in PS p 48:

$$u^s(p) = \begin{pmatrix} \sqrt{p.\sigma}\,\xi^s \\ \sqrt{p.\bar{\sigma}}\,\xi^s \end{pmatrix}, \tag{2.85}$$

and

$$v^s(p) = \begin{pmatrix} \sqrt{p.\sigma}\,\xi^s \\ -\sqrt{p.\bar{\sigma}}\,\xi^s \end{pmatrix}, \tag{2.86}$$

with, for instance, $\xi^1 = \begin{pmatrix} 1 \\ 0 \end{pmatrix}$ representing a particle with a spin up in the z-direction in the rest frame and $\xi^2 = \begin{pmatrix} 0 \\ 1 \end{pmatrix}$ representing a particle with a spin down in the z-direction in the rest frame.

Exercise 10: Check the explicit form of the solutions, their orthogonality relations and spin sums given in PS.

Note that the explicit form of the solutions clarifies the convention regarding the left and right spinors. For instance, if we consider a massless (or an ultra relativistic) particle moving in the z-direction, we have $p_0 = E$ and $p_3 = -E$ (because we lowered the index), and

$$\sqrt{p.\sigma} = \begin{pmatrix} 0 & 0 \\ 0 & \sqrt{2E} \end{pmatrix} \text{ and } \sqrt{p.\bar{\sigma}} = \begin{pmatrix} \sqrt{2E} & 0 \\ 0 & 0 \end{pmatrix}. \tag{2.87}$$

Consequently, for $s = 1$, the first two components are zero and the third and fourth components correspond to a 'right-handed' particle with its spin in the same direction as its momentum. For $s = 2$, the first two components correspond to a 'left-handed' particle with its spin opposite to the direction of the momentum and the third and fourth components are zero. This description can be summarized by using the *helicity* which is the projection of the spin on the direction of the momentum

$$h = \hat{p} \cdot \mathbf{S}. \tag{2.88}$$

Positive helicity corresponds to right-handed particles while negative helicity corresponds to left-handed particle (see PS p 47).

The Dirac Lagrangian has discrete symmetries that will be discussed more easily after performing the quantization. One of them is the parity transformation which for $D = 4$ reads

$$\Lambda_{\text{Parity}} = \begin{pmatrix} 1 & 0 & 0 & 0 \\ 0 & -1 & 0 & 0 \\ 0 & 0 & -1 & 0 \\ 0 & 0 & 0 & -1 \end{pmatrix}. \tag{2.89}$$

This matrix has a determinant equal to -1 and cannot be obtained by exponentiating traceless matrices. The corresponding $\Lambda_{1/2,\text{parity}} = \gamma^0$ satisfies the covariance condition (2.51) because its square is the identity and it anticommutes with the spatial γ^i. Notice that γ^0 swaps left and right spinors.

2.6 Maxwell fields

The manifestly relativistic description of Maxwell's equations is discussed at length in standard textbooks [7]. The basic object is the antisymmetric field-strength tensor

$$F_{\mu\nu} = \partial_\mu A_\nu - \partial_\nu A_\mu. \tag{2.90}$$

As numerical solutions of E and M problems are often carried in MKSA units, we will restore the factors of c corresponding to these units in this section. In the following we work in $D = 4$. The Lagrangian density is

$$\mathcal{L}^{\text{Maxwell}} = -(1/4)F_{\mu\nu}F^{\mu\nu} - \mu_0 J^\mu A_\mu. \tag{2.91}$$

Imposing $\delta S = 0$ yields Maxwell's equations with charge and currents

$$\partial_\mu F^{\mu\nu} = \mu_0 J^\nu, \tag{2.92}$$

if we identify $F^{0i} = -E^i/c$, $F^{ij} = -\epsilon^{ijk}B^k$, $J^0 = \rho c$, $A^0 = \phi/c$ while the three-vectors of J^μ and A^μ are the usual currents $\mathbf{J}$ and potentials $\mathbf{A}$ three-vectors.

Exercise 11: Check the variational result and the correspondence with Maxwell equations in MKSA.

Note that the equations of motion (2.92) and the identification of the electric field remain valid for any D. However, the identification of the magnetic field is D-dependent. For instance for $D = 2$, there is no magnetic field and for $D = 3$, it is just a parity odd (pseudo)scalar density. The homogeneous Maxwell equations are also D-dependent. For $D = 4$, we can define a *dual* tensor

$$\tilde{F}^{\mu\nu} \equiv \frac{1}{2}\epsilon^{\mu\nu\rho\sigma}F_{\rho\sigma} = \epsilon^{\mu\nu\rho\sigma}\partial_\rho A_\sigma, \tag{2.93}$$

where $\epsilon^{\mu\nu\rho\sigma}$ is the totally antisymmetric Levi-Civita tensor with the sign convention $\epsilon^{0123} = +1$. Each permutation of neighbor indices generates a change of sign, for instance $\epsilon^{1023} = -1$, $\epsilon^{1032} = +1$ etc. Since two gradients commute, the definition (2.93) implies that

$$\partial_\mu \tilde{F}^{\mu\nu} = 0. \tag{2.94}$$

Exercise 12: Check that this equation implies the homogeneous Maxwell equations.

Exercise 13: Following the $D = 4$ path, derive Maxwell equations for $D = 3$.

Use $\tilde{F}^\mu \equiv \frac{1}{2}\epsilon^{\mu\rho\sigma}F_{\rho\sigma}$ and $F^{ij} = -\epsilon^{ij}B$.

The description of the **E** and **B** fields in terms of A_μ is not one to one. The gauge transformation

$$A_\mu \to A_\mu - \partial_\mu\alpha, \tag{2.95}$$

leaves $F_{\mu\nu}$ unchanged. This means that some of the degrees of freedom of A_μ are unphysical but it also allows us to make a gauge transformation such that the transformed A_μ satisfy some 'gauge condition' that may simplify the problem under consideration. As we will discuss later, the temporal gauge $A_0 = 0$ is convenient in the Hamiltonian formalism but not Lorentz-invariant. For a manifestly relativistic formulation, the Lorenz gauge

$$\partial_\mu A^\mu = 0, \tag{2.96}$$

is Lorentz-invariant and plays a special role. Note that the two last names are very similar but different. The Lorenz gauge condition has a residual invariance: under a gauge transformation (2.95),

$$\partial_\mu A^\mu \to -\Box\alpha, \tag{2.97}$$

has no effect if α is a solution of the massless Klein–Gordon equation. In the Lorenz gauge, Maxwell equations with charges and currents are simply

$$\Box A^\nu = \mu_0 J^\nu. \tag{2.98}$$

The solutions of the sourceless Maxwell equations in the Lorenz gauge have the form

$$A^\mu(x) = \epsilon^\mu(p)\exp(-ip \cdot x), \tag{2.99}$$

with the condition $p_\mu p^\mu = 0$. The Lorenz gauge condition implies that

$$p_\mu \epsilon^\mu(p) = 0. \tag{2.100}$$

This allows us to express one component of ϵ^μ in terms of the other quantities present. The residual gauge symmetry associated with a gradient of a solution of the massless Klein–Gordon equation guarantees the additional gauge freedom:

$$\epsilon^\mu(p) \to \epsilon^\mu(p) + \lambda p^\mu, \tag{2.101}$$

for arbitrary λ. After using this additional freedom, we are left with $D - 2$ transverse polarizations. As an example in $D = 4$, if p^μ represents the motion in the z direction $(E, 0, 0, E)$, the polarization ϵ^μ can be linear combinations of $(1, 0, 0, 1)$, $(0, 1, 0, 0)$ and $(0, 0, 1, 0)$. The first possibility can be eliminated with the residual gauge transformation and we are left with two transverse polarizations ϵ^1 and ϵ^2. After this is done, we end up with a plane wave solution satisfying the conditions $A^0 = 0$ and $\nabla \cdot \mathbf{A} = 0$.

It is possible to add a gauge symmetry breaking term

$$\mathcal{L}^{gsb} = -\frac{\lambda}{2}(\partial_\mu A^\mu)^2. \tag{2.102}$$

With this extra term, the equations of motion become

$$\Box A^\nu + (\lambda - 1)\partial^\nu(\partial_\mu A^\mu) = \mu_0 J^\nu. \tag{2.103}$$

By picking $\lambda = 1$, we recover equation (2.98). This choice is called the 'Feynman gauge'.

2.7 Yang–Mills fields

In equation (2.75) we have constructed the covariant derivative of ψ denoted $D_\mu\psi$ which transforms in the same way as ψ under a local phase transformation. This construction can be extended to the case of a column of N spinors ψ^a that we denote Ψ. The a indices take values $1, \ldots N$ and we will refer to them as 'color indices' to distinguish them from global spinor or flavor indices. We consider a local transformation:

$$\Psi(x) \rightarrow V(x)\Psi(x), \tag{2.104}$$

where $V(x)$ is a unitary $N \times N$ matrix mixing the color indices and not affecting the spinor indices. We want to construct a covariant derivative $D_\mu\Psi$, such that under the same transformation, we have (with color indices implicit)

$$D_\mu\Psi(x) \rightarrow V(x)D_\mu\Psi(x). \tag{2.105}$$

Following what was done in equation (2.75) for a single spinor, we introduce gauge potentials A_μ which are now $N \times N$ matrices. If we want to make the color indices explicit:

$$D_\mu{}^a{}_b \equiv \partial_\mu\delta^a{}_b + iA_\mu(x)^a{}_b. \tag{2.106}$$

One can check that the transformation

$$A_\mu(x) \rightarrow V(x)A_\mu V^{-1}(x) - iV(x)(\partial_\mu V^{-1}(x)), \tag{2.107}$$

guarantees that the transformation is homogeneous as in equation (2.105). Note that in the above equation, the derivative ∂_μ acts only on V. In order to check that the inhomogeneous terms cancel, we used the identity

$$(\partial_\mu V)V^{-1} = -V(\partial_\mu V^{-1}). \tag{2.108}$$

We are in position to define the non-abelian generalization of the field-strength tensor which is now an $N \times N$ matrix:

$$F_{\mu\nu} = -i[D_\mu, D_\nu] = \partial_\mu A_\nu - \partial_\nu A_\mu + i[A_\mu, A_\nu], \tag{2.109}$$

which transforms homogeneously:

$$F_{\mu\nu} \rightarrow V F_{\mu\nu} V^{-1}. \tag{2.110}$$

Exercise 14: Check the details of all the equations for transformations in this subsection.

We can use the cyclicity of the trace of finite dimensional matrices $(\text{Tr}(ABC) = \text{Tr}(CAB) = \text{Tr}(BCA))$, which implies that $\text{Tr}(VAV^{-1}) = \text{Tr}\,A$ to construct a gauge-invariant (invariant under a local symmetry) quantities. From the homogeneous transformation (2.110), we see that

$$\mathcal{L}^{\text{Yang–Mills}} = -K\,\text{Tr}(F_{\mu\nu}F^{\mu\nu}), \tag{2.111}$$

is gauge invariant. The constant K depends on the normalization of the basis of matrices used to express $A_\mu(x)^a{}_b$. Since $i\partial_\mu$ is a Hermitian operator, in order to have $iD\mu$ Hermitian, we require that $A_\mu(x)^a{}_b$ be a Hermitian matrix and we express it as

$$A_\mu(x)^a{}_b = A_\mu^A(x)T^{A\,a}{}_b, \tag{2.112}$$

with $T^{A\,a}{}_b$ a basis of Hermitian matrices. If we impose that V has determinant 1 and that V is a group element of $SU(N)$, the $T^{A\,a}{}_b$, are traceless $N \times N$ matrices and the index A runs from 1 to $N^2 - 1$. This is a specific representation of $SU(N)$ called the fundamental representation. We often use the normalization

$$\text{Tr}(T^A T^B) = \frac{1}{2}\delta^{AB}, \tag{2.113}$$

for the fundamental representation. The construction can be extended to other representations and other groups.

Exercise 15: Calculate the equation of motion for $\mathcal{L}^{\text{Yang–Mills}}$.

2.8 Linear sigma models

Following references [8–11], we consider a $N_f \times N_f$ matrix of effective fields ϕ_{ij} having the same quantum numbers as $\bar{\psi}_{Rj}\psi_{Li}$ with the summation over the color indices implicit. Under a general transformation of $U(N_f)_L \otimes U(N_f)_R$, we have

$$\phi \to U_L \phi U_R^\dagger. \tag{2.114}$$

We now use a basis of $N_f \times N_f$ Hermitian matrices Γ^α such that

$$\text{Tr}(\Gamma^\alpha \Gamma^\beta) = (1/2)\delta^{\alpha\beta}, \tag{2.115}$$

to express ϕ in terms of N_f^2 scalars (0^+ in J^P notation), denoted S_α, and N_f^2 pseudoscalars (0^-), denoted P_α:

$$\phi = (S_\alpha + iP_\alpha)\Gamma^\alpha, \tag{2.116}$$

with a summation over $\alpha = 0, 1, \ldots N_f^2 - 1$. We use the convention that $\Gamma_0 = \mathbb{1}/\sqrt{2N_f}$ while the remaining $N_f^2 - 1$ matrices are traceless.

We introduce the diagonal subgroup $U(N_f)_V$ defined by the elements of $U(N_f)_L \otimes U(N_f)_R$ such that $U_L = U_R$. Using equations (2.116) and (2.114), we see that under $U(N_f)_V$, S_0 and P_0 are singlets denoted σ and η' respectively, while the remaining components transform like the adjoint representation and are denoted $\mathbf{a}_0$ and $\boldsymbol{\pi}$, respectively.

We follow the notations used in reference [12] and consider the effective Lagrangian

$$\mathcal{L} = \text{Tr}\partial_\mu\phi\partial^\mu\phi^\dagger - V, \tag{2.117}$$

with the potential split into three parts

$$V = V_0 + V_a + V_m, \tag{2.118}$$

that we now proceed to define and discuss. The first term is the most general $U(N_f)_L \otimes U(N_f)_R$ invariant renormalizable expression:

$$V_0 \equiv -\mu^2 \, \text{Tr}(\phi^\dagger\phi) + (1/2)(\lambda_\sigma - \lambda_{a0})(\text{Tr}(\phi^\dagger\phi))^2$$
$$+ (N_f/2)\lambda_{a0} \, \text{Tr}((\phi^\dagger\phi)^2). \tag{2.119}$$

The use of $\lambda_\sigma - \lambda_{a0}$ will become clear when we write the mass formulas. The stability of V_0 is discussed in [12]. Note that first two terms and the kinetic term have a larger group of symmetry $O(2N_f^2)$. The second term

$$V_a \equiv -2(2N_f)^{N_f/2-2}X(det\phi + det\phi^\dagger), \tag{2.120}$$

is invariant under $SU(N_f)_L \otimes SU(N_f)_R$ but breaks the axial $U(1)_A$. It takes into account the effect of the axial anomaly for the fundamental representation. Finally, the third term represents the effect of mass term which is the same for the N_f flavors:

$$V_m \equiv -(b/\sqrt{2N_f})(\text{Tr} \, \phi + \text{Tr} \, \phi^\dagger) = -b\sigma. \tag{2.121}$$

It is invariant under $SU(N)_V$.

In the following, we assume that chiral symmetry is spontaneously broken by an $SU(N_f)$-invariant vacuum expectation value (v.e.v.):

$$\langle\phi_{ij}\rangle = v\delta_{ij}/\sqrt{2N_f}. \tag{2.122}$$

This amounts to saying that $\langle\sigma\rangle = v$ while the other v.e.v.s are zero. The other results for the spectrum can be written in a compact way:

$$M_{\eta'}^2 - M_\pi^2 = Xv^{N_f-2}$$
$$M_\sigma^2 - M_\pi^2 = \lambda_\sigma v^2 - (1 - 2/N_f)Xv^{N_f-2} \tag{2.123}$$
$$M_{a0}^2 - M_\pi^2 = \lambda_{a0}v^2 + (2/N_f)Xv^{N_f-2}.$$

2.9 General relativity

The notions of covariance, covariant derivative and the idea of taking the commutator of two covariant derivatives were developed long before the Yang–Mills construction by mathematicians studying Riemannian geometry. A central object in this approach is the *metric*. So far we have worked in a 'flat' space–time and used a constant metric $g_{\mu\nu}$. In a curved space–time, the metric becomes non-trivial. This will allow us to write classical field theory in curved background in a way that is invariant under general coordinate transformations $x^\mu(x')$.

In the following, we will consider curved spaces which can be embedded in a flat space with more dimensions. A well-known example is the n-dimensional sphere S_n, which can be embedded in an $n + 1$ dimensional Euclidean space. More generally, we consider flat coordinates

$$X^A, \ A = 1, 2, \ ...n, \ n + q. \tag{2.124}$$

and a constant metric η_{AB}. We follow the simple presentation of Dirac [13] in a course on general relativity intended for undergraduate students. In section 2.10 and appendix B, we discuss the case $q = 1$ and special cases where the metric has only ± 1 elements. In addition to S_n, this includes curved backgrounds such as the de Sitter (dS_n) and anti-de Sitter (AdS_n) spaces.

The curved space has coordinates

$$x^\mu, \ \mu = 1, 2, \ ...n. \tag{2.125}$$

and its embedded elements (points) are denoted $X^A(x)$. Arbitrary tangent vectors can be expressed in the $n + q$-dimensional space by using linear combinations of

$$t_\mu^A = \frac{\partial X^A(x)}{\partial x^\mu}. \tag{2.126}$$

If we use different curvilinear coordinates x'^μ and re-express $x^\mu(x')$, then

$$\frac{\partial X^A(x(x'))}{\partial x'^\mu} = \frac{\partial X^A(x(x'))}{\partial x^\nu}\frac{\partial x^\nu}{\partial x'^\mu}, \tag{2.127}$$

transforms like a *covariant vector*. We can now look at an arbitrary vector field $v^\mu(x)$ expressed locally in terms of tangent vectors. In the embedding space, it can be expressed as

$$V^A(x) = v^\mu(x)\frac{\partial X^A}{\partial x^\mu}. \tag{2.128}$$

For a given point of the curved space, its expression is independent of the curvilinear coordinates. This implies that $v^\mu(x)$ transforms like a contravariant vector. We now define the curved metric with two covariant indices:

$$g_{\mu\nu} \equiv \eta_{AB}\frac{\partial X^A}{\partial x^\mu}\frac{\partial X^B}{\partial x^\nu}. \tag{2.129}$$

As in special relativity (2.23), we can convert a contravariant vector field into a covariant one by lowering the index with the metric:

$$v_\mu(x) = g_{\mu\nu}(x)v^\nu(x) = \eta_{AB}\frac{\partial X^A}{\partial x^\mu}V^B. \tag{2.130}$$

If we 'parallel transport' v_μ, keeping V^B unchanged, from x^ν to $x^\nu + dx^\nu$, the change in coordinates for the covariant vector fields is

$$
\begin{aligned}
dv_\mu &= v_\mu(x^\nu + dx^\nu) - v_\mu(x^\nu) = \eta_{AB}\frac{\partial^2 X^A}{\partial x^\mu \partial x^\nu}dx^\nu V^B \\
&= \eta_{AB}\frac{\partial^2 X^A}{\partial x^\mu \partial x^\nu}dx^\nu v^\rho\frac{\partial X^B}{\partial x^\rho}.
\end{aligned}
\tag{2.131}
$$

Defining the Christoffel symbols of the first kind $\Gamma_{\rho\mu\nu}$ by the formula

$$dv_\mu = v^\rho\Gamma_{\rho\mu\nu}dx^\nu, \tag{2.132}$$

we obtain the explicit expression

$$\Gamma_{\rho\mu\nu} = \eta_{AB}\frac{\partial^2 X^A}{\partial x^\mu \partial x^\nu}\frac{\partial X^B}{\partial x^\rho}, \tag{2.133}$$

which is clearly symmetric in its last two indices μ and ν.

Exercise 16: Show that

$$\Gamma_{\rho\mu\nu} = \frac{1}{2}(g_{\rho\mu,\nu} + g_{\rho\nu,\mu} - g_{\mu\nu,\rho}), \tag{2.134}$$

with $\ldots,_\mu$ a short notation for $\frac{\partial \ldots}{\partial x^\mu}$.

Solution. See reference [13].

Under a change in coordinates $x^\mu(x')$,

$$\frac{\partial^2 X^A}{\partial x^\mu \partial x^\nu} \to \frac{\partial^2 X^A(x(x'))}{\partial x'^\mu \partial x'^\nu} = \frac{\partial^2 X^A(x)}{\partial x^\rho \partial x^\sigma}\frac{\partial x^\rho}{\partial x'^\mu}\frac{\partial x^\sigma}{\partial x'^\nu} + \frac{\partial X^A(x)}{\partial x^\sigma}\frac{\partial^2 x^\sigma}{\partial x'^\mu \partial x'^\nu}. \tag{2.135}$$

The first term corresponds to the normal transformation of a rank-2 covariant tensor, but the second term is an inhomogeneous term reminiscent of the transformation of gauge fields. A similar inhomogeneous term appears in the transformation of the derivative of a vector field

$$\frac{\partial v_\nu(x)}{\partial x^\mu} \to \frac{\partial}{\partial x'^\mu}\left[v_\sigma(x(x'))\frac{\partial x^\sigma}{\partial x'^\nu}\right] = \frac{\partial v_\sigma(x)}{\partial x^\rho}\frac{\partial x^\rho}{\partial x'^\mu}\frac{\partial x^\sigma}{\partial x'^\nu} + v_\sigma\frac{\partial^2 x^\sigma}{\partial x'^\mu \partial x'^\nu}. \tag{2.136}$$

We can also raise indices using

$$v^\mu = g^{\mu\nu}v_\mu, \tag{2.137}$$

provided that

$$g^{\mu\nu}g_{\nu\rho} = \delta^\mu_\rho, \tag{2.138}$$

With this we can define the Christoffel symbols of the second kind

$$\Gamma^\rho_{\mu\nu} \equiv g^{\rho\sigma}\Gamma_{\sigma\mu\nu}. \tag{2.139}$$

This allows us to write

$$dv_\mu = \Gamma^\rho_{\mu\nu}v_\rho dx^\nu. \tag{2.140}$$

It is clear that

$$v_\mu w^\mu = v_\mu g^{\mu\nu}w_\nu, \tag{2.141}$$

is coordinate invariant and does not change under parallel transport. Consequently,

$$d(v_\mu w^\mu) = dv_\mu w^\mu + v_\mu dw^\mu = 0. \tag{2.142}$$

From equation (2.140), this implies

$$v_\mu dw^\mu = -w^\mu \Gamma^\rho_{\mu\nu}v_\rho dx^\nu, \tag{2.143}$$

and also

$$dw^\mu = -\Gamma^\mu_{\sigma\nu}w^\sigma dx^\nu, \tag{2.144}$$

Exercise 17: Calculate the tangent vectors, the metric, the Christoffel symbols for a two-dimensional sphere. Discuss the geometrical interpretation of the Christoffel symbols of the second kind.

Solution. We use the standard spherical coordinates x^μ, with $x^1 = \phi$ and $x^2 = \theta$, for a sphere of radius 1 embedded in a three-dimensional Euclidean space: $X^1 = \sin\theta\cos\phi$, $X^2 = \sin\theta\sin\phi$, $X^3 = \cos\theta$. We call the basis tangent vectors $\partial\vec{X}/\partial x^\mu \equiv \vec{t}_\mu$. The $\vec{t}_\theta$ are along meridians and have a constant length 1. The $\vec{t}_\phi$ are tangent to circles which are parallels of constant θ and radius $\sin\theta$ which is also the length of $\vec{t}_\phi$. These shrink to zero as we reach the north or south poles. The two types of tangent vectors are orthogonal. We conclude that $g_{\theta\theta} = g^{\theta\theta} = 1$ and $g_{\phi\phi} = \sin\theta^2 = 1/g^{\phi\phi}$. The off-diagonal elements are zero. Using equation (2.134), we obtain

$$\Gamma_{\theta\phi\phi} = -\sin\theta\cos\theta, \text{ and } \Gamma_{\phi\phi\theta} = \Gamma_{\phi\theta\phi} = \sin\theta\cos\theta. \tag{2.145}$$

The θ indices can be raised freely while ϕ raising requires a factor $1/\sin\theta^2$. Remembering that the contravariant changes have a minus, we obtain

$$dA^\theta = \sin\theta\cos\theta A^\phi d\phi, \tag{2.146}$$

$$dA^\phi = -\cot\theta(A^\phi d\theta + A^\theta d\phi). \tag{2.147}$$

The first line means that if a tangent vector in the ϕ direction, $A^\phi \vec{t}_\phi$, is transported parallel to itself along a parallel, it has a length $A^\phi \sin\theta$ and develops a component normal to the parallel circle in the plane of constant θ. This vector has a length $A^\phi \sin\theta d\phi$. Seen in the three-dimensional embedding space, it has a component normal to the sphere and a component tangent in the θ direction which costs a factor $\cos\theta$. This combines into $\sin\theta \cos\theta A^\phi d\phi$ as expected.

If the same vector is transported in the θ direction, it stays tangent in the ϕ direction while keeping its length. The length of $\vec{t}_\phi$ increases by a factor $\cos\theta d\theta$. To maintain the length we need to reduce A^ϕ, more precisely

$$(A^\phi + dA^\phi)(\sin\theta + \cos\theta d\theta) = A^\phi \sin\theta, \tag{2.148}$$

which implies $dA^\phi = -\cot\theta A^\phi d\theta$.

Third, if a tangent vector $A^\theta \vec{t}_\theta$ is transported along a parallel by $d\phi$, it will develop a negative component proportional to $d\phi$ along $\vec{t}_\phi$. Two meridians separated by $d\phi$ have a relative angle $d\phi$ near the poles and are parallel near the equator so the effect is proportional to $\cos\theta$. Putting everything together, we get

$$dA^\phi \sin\theta = -\cos\theta A^\theta d\phi, \tag{2.149}$$

in agreement with the result obtained from the analytical formula.

We can now write covariant derivatives

$$\mathfrak{D}_\mu v_\nu \equiv \partial_\mu v_\nu - \Gamma^\rho_{\mu\nu} v_\rho, \tag{2.150}$$

and

$$\mathfrak{D}_\mu v^\nu \equiv \partial_\mu v^\nu + \Gamma^\nu_{\mu\rho} v^\rho, \tag{2.151}$$

which transform as rank two tensors.

Exercise 18: Check the transformation properties of the covariant derivatives above.

Solution. The inhomogeneous terms appearing in equations (2.135) and (2.136) cancel.

The covariant derivative of higher order tensor can be constructed in a similar way, with one Christoffel term per index and a minus (plus) sign for covariant (contravariant) indices. We are now in position to conclude our discussion of the analogy between Riemannian geometry and Yang–Mills theories. In curved space, the covariant derivatives do not commute. This is encoded in the Riemann tensor defined as

$$[\mathfrak{D}_\mu, \mathfrak{D}_\nu] v_\rho = R^\sigma_{\rho\mu\nu} v_\sigma. \tag{2.152}$$

Exercise 19: Using the explicit form of the covariant derivatives in terms of the Christoffel symbols, show that

$$R^\sigma_{\rho\mu\nu} = \Gamma^\sigma_{\rho\nu,\mu} - \Gamma^\sigma_{\rho\mu,\nu} + \Gamma^\alpha_{\rho\nu}\Gamma^\sigma_{\alpha\mu} - \Gamma^\alpha_{\rho\mu}\Gamma^\sigma_{\alpha\nu}. \tag{2.153}$$

Solution. See chapter 11 of [13]. The analogy with equation (2.109) is clear if we replace the color indices by space–time indices.

Finally, we define the Ricci tensor

$$R_{\rho\mu} = R^{\nu}_{\rho\mu\nu}, \tag{2.154}$$

and the curvature tensor

$$R = g^{\rho\mu}R_{\rho\mu}, \tag{2.155}$$

which appear in the equation of motion for gravitational theories.

Actions invariant under general coordinate transformations can be constructed by inserting a factor $\sqrt{|g|}$ beside $d^D x$, where g means the determinant of the metric, in space–time integrals of quantities that are invariant under general coordinate transformations. A well-known example is the Einstein–Hilbert action

$$S_{EH} = C \int d^4x \sqrt{|g|}\, R. \tag{2.156}$$

We can also combine matter fields with Lorentz indices and their covariant derivatives in a coordinate invariant way.

***Exercise 20*:** Show that varying the metric in the Einstein–Hilbert action leads to the equation of motion

$$R^{\mu\nu} - \frac{1}{2}g^{\mu\nu}R = 0. \tag{2.157}$$

Solution. See chapter 26 of [13].

An interesting application of the variational calculus is the determination of geodesics which are the curves of minimal proper length joining two given points. The proper length to minimize is

$$S_G = \int \sqrt{g_{\mu\nu}dx^\mu dx^\nu}, \tag{2.158}$$

along a curve $x^\mu(s)$ parametrized by s and joining $x_i^\mu = x^\mu(s_i)$ and $x_f^\mu = x^\mu(s_f)$, we can write

$$S_G = \int_{s_i}^{s_f} ds \sqrt{g_{\mu\nu}\frac{dx^\mu}{ds}\frac{dx^\nu}{ds}}. \tag{2.159}$$

We can adapt the variational methods of classical mechanics with a 'Lagrangian' $\sqrt{g_{\mu\nu}\frac{dx^\mu}{ds}\frac{dx^\nu}{ds}}$ and a 'time' s to show that the so-called geodesic equation

$$\ddot{x}^\rho + \Gamma^\rho_{\mu\nu}\dot{x}^\mu\dot{x}^\nu = 0, \tag{2.160}$$

provides a solution to $\delta S_G = 0$. The dot means derivation with respect to s.

Exercise 21: Derive this equation.

Solution. If we define

$$\tilde{L} \equiv g_{\mu\nu}\dot{x}^{\mu}\dot{x}^{\nu}, \tag{2.161}$$

$$\left(\frac{d}{ds}\frac{\partial}{\partial\dot{x}^{\sigma}} - \frac{\partial}{\partial x^{\sigma}}\right)\tilde{L} = g_{\sigma\nu}\ddot{x}^{\nu} + g_{\sigma\nu,\mu}\dot{x}^{\mu}\dot{x}^{\nu} - \frac{1}{2}g_{\mu\nu,\sigma}\dot{x}^{\mu}\dot{x}^{\nu}. \tag{2.162}$$

Symmetrizing on μ and ν in the second term, using equation (2.134) and raising σ with the metric we already obtain the geodesic equation (2.160). If we now consider the Euler–Lagrange equation for $F(\tilde{L})$:

$$\left(\frac{d}{ds}\frac{\partial}{\partial\dot{x}^{\sigma}} - \frac{\partial}{\partial x^{\sigma}}\right)F(\tilde{L}) = F'\left(\frac{d}{ds}\frac{\partial}{\partial\dot{x}^{\sigma}} - \frac{\partial}{\partial x^{\sigma}}\right)\tilde{L} + F''\frac{\partial\tilde{L}}{\partial\dot{x}^{\sigma}}\frac{d\tilde{L}}{ds}. \tag{2.163}$$

We can then show that $\frac{d\tilde{L}}{ds}$ is equation (2.160) contracted with $\dot{x}_{\rho}$. Taking $F(\tilde{L}) = \sqrt{\tilde{L}}$ we reach the desired conclusion.

2.10 Examples of two-dimensional curved spaces

In the following we discuss examples of two-dimensional curved spaces defined by a quadratic form of three flat coordinates X^1, X^2 and X^3. The generalizations to higher dimensions are discussed in appendix B. These include homogeneous and isotropic space–times relevant for cosmology. A nice introduction to this subject can be found in reference [14].

For the well-known example of the two-dimensional sphere S_2, we use the flat Euclidean metric η_{AB} with a signature $(+, +, +)$. A sphere of radius R is defined by

$$X^A X^B \eta_{AB} = (X^1)^2 + (X^2)^2 + (X^3)^2 = R^2. \tag{2.164}$$

The hyperbolic space $\mathbb{H}_2$ is defined as the solutions of

$$X^A X^B \eta_{AB} = (X^1)^2 + (X^2)^2 - (X^3)^2 = -R^2, \tag{2.165}$$

say with $X_3 > 0$. In the language of special relativity, the solutions of the quadratic equation are two disconnected time-like hyperboloids. Here we pick the one corresponding to the future.

de Sitter spaces are often discussed in the context of cosmology. For dS_2 we use a metric with signature $(+, +, -)$. The space is defined as the solutions of

$$X^A X^B \eta_{AB} = (X^1)^2 + (X^2)^2 - (X^3)^2 = R^2. \tag{2.166}$$

In the language of special relativity, the solutions of the quadratic equation form a single space-like hyperboloid. The anti-de Sitter space AdS_2 is defined by

$$X^A X^B \eta_{AB} = (X^1)^2 - (X^2)^2 - (X^3)^2 = -R^2. \tag{2.167}$$

In two dimensions this is equivalent to dS_2 with the order of the variables flipped. For $n \geqslant 3$, dS_n and AdS_n are not equivalent (see appendix B).

Exercise 22: Discuss the solutions of the geodesic equation for H_2 and dS_2. Represent typical solutions graphically in the three-dimensional space.

Solution. See appendix B where it is shown that the geodesics are confined to a plane going through the origin and containing the initial position and velocity. This also applies to the sphere and results in great circles, like the meridians or the equator. The solutions can also be constructed analytically by solving linear differential equations. The results are illustrated in figures 2.1–2.5. Mathematica notebooks to make some of these figures are appended.

Exercise 23: Map the coordinates X^1 and X^2 of H_2 into the open Poincaré disk using $X^1 = \sinh\chi \cos\phi$, $X^2 = \sinh\chi \sin\phi$ and set $\sinh\chi = 2r/(1 - r^2)$. Show graphically that the geodesics of H_2 are mapped into arcs of circle attached to the open boundary.

Solution. See figure 2.2 and the appended Mathematica notebook.

Figure 2.1. A geodesic on H_2 at the intersection of the space and a plane containing initial position and velocity (seen from two different angles).

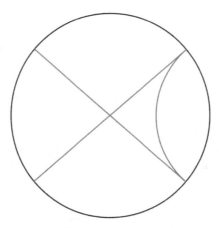

Figure 2.2. The same geodesic on H_2 but with a larger proper length (blue), mapped into the Poincaré disk. The red lines represent the asymptotic behavior (the angle is the same as in H_2).

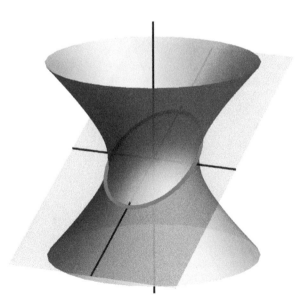

Figure 2.3. Two closed geodesics on dS_2 at the intersection of the space and a plane containing initial position and velocity.

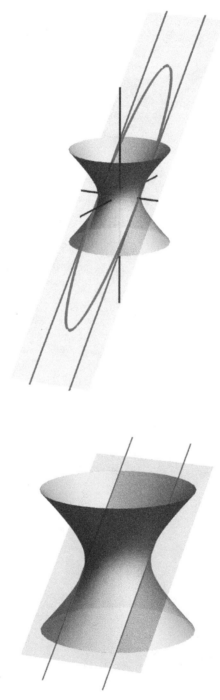

Figure 2.4. One closed geodesics on dS_2 at the intersection of the space and a plane containing initial position and velocity. The inclination of the plane is very close to 45° and the ellipse is getting elongated (top, blue). When the inclination is exactly 45°, the ellipse turns into two straight lines (red, top and bottom).

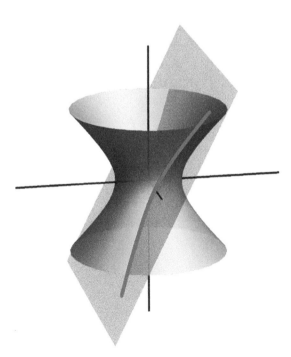

Figure 2.5. One open geodesic on dS_2 at the intersection of the space and a plane containing initial position and velocity (seen from different angles). The plane is tilted by more than $45°$ and the geodesic can extend to infinity.

2.11 Mathematica notebook for geodesics

```
In[1]:= (*H_2 parametrization*)
        XV[chi_, phi_] := {Sinh[chi] * Cos[phi], Sinh[chi] * Sin[phi], Cosh[chi]};
        (*tangent vectors*)
        dc = D[XV[chi, phi], chi];
        dp = D[XV[chi, phi], phi];
        (*embedding metric*)
        Eta = {{1, 0, 0}, {0, 1, 0}, {0, 0, -1}};
        (*curved space metric*)
        TableForm[{{Simplify[dc.Eta.dc], Simplify[dc.Eta.dp]},
          {Simplify[dp.Eta.dc], Simplify[dp.Eta.dp]}}]
```

```
Out[5]//TableForm=
        1      0
        0      Sinh[chi]^2
```

```
In[ ]:= (*tangent vectors in chi and phi directions*)
        {dc, dp}
```

```
Out[ ]= {{Cos[phi] Cosh[chi], Cosh[chi] Sin[phi], Sinh[chi]},
          {-Sin[phi] Sinh[chi], Cos[phi] Sinh[chi], 0}}
```

```
In[6]:= (*part of the hyperbolic space*)
        p1 = ParametricPlot3D[XV[chi, phi], {chi, -2., 2.}, {phi, 0, 2 * Pi}, Mesh → False,
          Axes -> False, Boxed → False, PlotStyle → {Opacity[0.9]}, PlotRange → All]
```

Out[6]=

```
In[7]:= (*geodesics*)
    (*initial data*)
    chi0 = 1.;
    phi0 = 0;
    chidot0 = 0;
    phidot0 = 1;
    (*initial embedding coordinates and velocities*)
    X0 = XV[chi0, phi0];
    Xdot0 =
        (dc /. {chi → chi0, phi → phi0}) * chidot0 + (dp /. {chi → chi0, phi → phi0}) * phidot0;
    (* lambda and other quantities from geodesic equation solution*)
    lamb = Xdot0.Eta.Xdot0;
    XP = (1/2) * (X0 + (1/Sqrt[lamb]) * Xdot0);
    XM = (1/2) * (X0 - (1/Sqrt[lamb]) * Xdot0);
    (*geodesic solution*)
    geo[s_] := XP * Exp[Sqrt[lamb] * s] + XM * Exp[-Sqrt[lamb] * s];
    ge1 = ParametricPlot3D[Re[geo[s]], {s, -1.5, 1.5}, Boxed → False,
        Axes → False, PlotStyle → {RGBColor[0, 0, 1], Thickness[0.01]}]
```

Out[17]=

In[18]:= `(*plane with initial position and velocity*)`
```
plxy = ParametricPlot3D[s * Xdot0 + t * X0, {s, -5, 5}, {t, -1, 3}, Mesh → False,
  Axes → False, Boxed → False, PlotStyle → {Opacity[0.3], RGBColor[0, 0, 1]}]
```

Out[18]=

In[20]:=
```
px = ParametricPlot3D[{s, 0, 0}, {s, -3, 3}, PlotStyle → RGBColor[0, 0, 0]];
py = ParametricPlot3D[{0, s, 0}, {s, -5, 5}, PlotStyle → RGBColor[0, 0, 0]];
pz = ParametricPlot3D[{0, 0, s}, {s, -1, 10}, PlotStyle → RGBColor[0, 0, 0]];
```

In[23]:=
```
p1 = ParametricPlot3D[XV[chi, phi], {chi, -2., 2.}, {phi, 0, 2 * Pi}, Mesh → False,
  Axes -> False, Boxed → False, PlotStyle → {Opacity[0.9]}, PlotRange → All];
ge1 = ParametricPlot3D[Re[geo[s]], {s, -1.5, 1.5}, Boxed → False,
  Axes → False, PlotStyle → {RGBColor[0, 0, 1], Thickness[0.01]}];
pl = ParametricPlot3D[s * Xdot0 + t * X0, {s, -5, 5}, {t, -1, 3}, Mesh → False,
  Axes -> False, Boxed → False, PlotStyle → {Opacity[0.3], RGBColor[0, 0, 1]}];
px = ParametricPlot3D[{s, 0, 0}, {s, -3, 3}, PlotStyle → RGBColor[0, 0, 0]];
py = ParametricPlot3D[{0, s, 0}, {s, -5, 5}, PlotStyle → RGBColor[0, 0, 0]];
pz = ParametricPlot3D[{0, 0, s}, {s, -1, 10}, PlotStyle → RGBColor[0, 0, 0]];

sho = Show[ge1, p1, pl, px, py, pz, ViewPoint → {5, -10, 3}]
```

Out[28]=

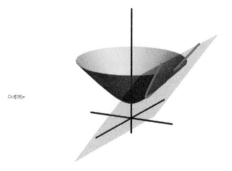

2-34

In[29]:=

```
(*conversion to polar coordinates*)
polarangle[x_, y_] := Module[
    (* private variable with name making their maning obvious*)
    {cos, sine, r, angle},
    r = Sqrt[x^2 + y^2];
    cos = x / r;
    sine = y / r;
    (*mutually exclusive cases*)
    If[cos == 0 && sine == 0, angle = "undefined"];
    If[cos == 0 && sine > 0, angle = Pi / 2];
    If[cos == 0 && sine < 0, angle = -Pi / 2];
    If[cos > 0, angle = ArcTan[sine / cos]];
    If[cos < 0 && sine > 0, angle = ArcTan[sine / cos] + Pi];
    If[cos < 0 && sine < 0, angle = ArcTan[sine / cos] - Pi];
    Return[angle]]

(*map into poincare disk*)
poincarer[s_] := (Sqrt[s^2 + 1] - 1) / s;
```

In[31]:= unitc =
```
ParametricPlot[{Cos[ph], Sin[ph]}, {ph, 0, 2 * Pi}, PlotStyle → RGBColor[0, 0, 0]]
```

Out[31]=

```
In[34]:= gepar2 =
    ParametricPlot[{poincarer[Sqrt[(Re[geo[s][[1]]])^2 + (Re[geo[s][[2]]])^2]]
        Cos[polarangle[Re[geo[s][[1]]], Re[geo[s][[2]]]]],
        poincarer[Sqrt[(Re[geo[s][[1]]])^2 + (Re[geo[s][[2]]])^2]] *
        Sin[polarangle[Re[geo[s][[1]]], Re[geo[s][[2]]]]]},
        {s, -5, 5}, Axes → False, PlotStyle → RGBColor[0, 0, 1]]

    Show[gepar2, unitc, PlotRange → All]
```

Out[34]=

```
In[36]:= asymp1 = ParametricPlot[{xx, -xx / Sinh[chi0]},
        {xx, -0.76, 0.76}, PlotStyle → RGBColor[1, 0, 0]];
    asymp2 = ParametricPlot[{xx, xx / Sinh[chi0]}, {xx, -0.76, 0.76},
        PlotStyle → RGBColor[1, 0, 0]];
```

In[38]:= `sho = Show[gepar2, unitc, asymp1, asymp2, PlotRange → All]`

Out[38]=

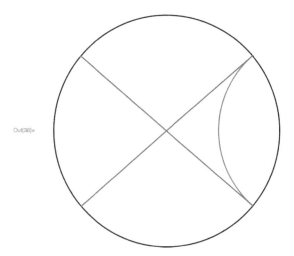

References

[1] Landau L D and Lifshitz E M 1976 *Mechanics Course of Theoretical Physics* vol 1 3rd edn (Oxford: Butterworth-Heinemann)

[2] Landau L D and Lifshitz E M 1980 *The Classical Theory of Fields* 4th edn (Oxford: Butterworth-Heinemann)

[3] Abers E S and Lee B W 1973 *Phys. Rep.* **9** 1–141

[4] Noether E 1971 *Transport Theor. Stat. Phys.* **1** 186–207

[5] Arnold V I 1973 *Ordinary Differential equations* (Cambridge, MA: MIT Press)

[6] Gelfand I, Minlos R and Cummins G 2012 *Representations of the Rotation and Lorentz Groups and Their Applications* (Eastford, CT: Martino Fine Books)

[7] Jackson J D 1998 *Classical Electrodynamics* (New York: Wiley)

[8] Schechter J and Ueda Y 1971 *Phys. Rev.* D **3** 2874–93

[9] Rosenzweig C, Schechter J and Trahern C G 1980 *Phys. Rev.* D **21** 3388–92

[10] 't Hooft G 1986 *Phys. Rep.* **142** 357–87

[11] Meurice Y 1987 *Mod. Phys. Lett.* **699** 699

[12] Meurice Y 2017 *Phys. Rev.* D **96** 114507

[13] Dirac P 1996 *General Theory of Relativity* (Princeton, NJ: Princeton University Press)

[14] Bengtsson I 1998 *Anti-de Sitter Space* https://web.archive.org/web/20180319192552/http://www.fysik.su.se/~ingemar/Kurs.pdf

Chapter 3

Canonical quantization

3.1 A one-dimensional harmonic crystal

We consider a one-dimensional harmonic crystal with N_s sites and Lagrangian

$$L = \frac{1}{2} \sum_{j=1}^{N_s} \left(\dot{q}_j^2 - A(q_{j+1} - q_j)^2 - Bq_j^2 \right),$$ (3.1)

with periodic boundary conditions $q_{N_s+1} = q_1$ (ring). There are N_s Euler–Lagrange equations of motion

$$\ddot{q}_j = A(q_{j+1} - 2q_j + q_{j-1}) - Bq_j.$$ (3.2)

Their solutions can be found by using a plane wave ansatz

$$q_j(k_l) = \exp(-i(\omega_l t - k_l j)),$$ (3.3)

where

$$k_l = \frac{2\pi}{N_s} l,$$ (3.4)

and l is an integer in order to enforce the periodic boundary conditions. Since j is an integer, shifting l by N_s has no effect on q_j and there are only N_s solutions. The result is

$$\omega_l^2 = B + 2A(1 - \cos(k_l)).$$ (3.5)

Notice that $\omega_{N_s-l}^2 = \omega_l^2$. For small k_l, we can expand the cosine in series and obtain

$$\omega_l^2 - Ak_l^2 \simeq B,$$ (3.6)

doi:10.1088/978-0-7503-2187-7ch3

which is reminiscent of the relativistic mass-shell condition in $1 + 1$ dimensions:

$$E^2 - p^2 = m^2, \tag{3.7}$$

written in $c = 1$ units. We can now use the Hamiltonian formalism to discuss the quantization of the model. We call p_j the canonical conjugate momentum of q_j and it is simply

$$p_j = \frac{\partial L}{\partial \dot{q}_j} = \dot{q}_j. \tag{3.8}$$

Following equation (2.7), the Hamiltonian reads

$$H = \frac{1}{2} \sum_{j=1}^{N_s} \left(p_j^2 + A(q_{j+1} - q_j)^2 + Bq_j^2 \right), \tag{3.9}$$

Hamilton equations read

$$\dot{q}_j = \frac{\partial H}{\partial p_j} = p_j, \tag{3.10}$$

and

$$\dot{p}_j = -\frac{\partial H}{\partial q_j} = A(q_{j+1} - 2q_j + q_{j-1}) - Bq_j. \tag{3.11}$$

We can quantize this model by assuming canonical commutation relations (in $\hbar = 1$ units):

$$[\hat{q}_j, \hat{p}_l] = i\delta_{jl} \text{ and } [\hat{q}_j, \hat{q}_l] = [\hat{p}_j, \hat{p}_l] = 0. \tag{3.12}$$

These equations define a quantum Hamiltonian that can be diagonalized by combining the classical solution discussed above and the standard creation/annihilation operators.

The solution of the quantum problem involves prefactors and ordering aspects that we shall first review in the well-known case of a single harmonic oscillator with Hamiltonian

$$H = \frac{1}{2}p^2 + \frac{1}{2}\omega^2 q^2. \tag{3.13}$$

The classical solutions of Hamilton equations can be obtained from the time evolution of

$$a \equiv \sqrt{\frac{\omega}{2}} \left(q + i\frac{p}{\omega} \right), \tag{3.14}$$

which satisfies a single complex linear equation

$$\dot{a} = -i\omega a. \tag{3.15}$$

The classical problem is then completely solved by using

$$a(t) = a(0) \exp(-i\omega t). \tag{3.16}$$

The spectrum of the quantum Hamiltonian

$$\hat{H} = \frac{1}{2}\hat{p}^2 + \frac{1}{2}\omega^2\hat{q}^2, \tag{3.17}$$

given the commutation relation $[\hat{q}, \hat{p}] = i$ is obtained by a similar algebraic method. The annihilation operator

$$\hat{a} \equiv \sqrt{\frac{\omega}{2}}\left(\hat{q} + i\frac{\hat{p}}{\omega}\right), \tag{3.18}$$

satisfies the commutation relation

$$[\hat{a}, \hat{a}^\dagger] = 1. \tag{3.19}$$

The prefactor in equation (3.18) has been chosen so that the commutator is exactly 1. We can now use

$$\hat{q} = \frac{1}{\sqrt{2\omega}}(\hat{a} + \hat{a}^\dagger), \text{ and } \hat{p} = -i\sqrt{\frac{\omega}{2}}(\hat{a} - \hat{a}^\dagger), \tag{3.20}$$

to express the Hamiltonian as

$$\hat{H} = \frac{1}{2}\omega(2\hat{a}^\dagger\hat{a} + 1). \tag{3.21}$$

The spectrum $E_n = \omega\left(n + \frac{1}{2}\right)$ follows from the standard representation of the commutation relation (3.19)

$$\hat{a}|0\rangle = 0, \quad \hat{a}^\dagger|n\rangle = \sqrt{n+1}\,|n+1\rangle, \quad \hat{a}|n\rangle = \sqrt{n}\,|n-1\rangle. \tag{3.22}$$

As a mnemonic notice that the largest integer in the two states appears in the square root.

A similar procedure can be followed for our problem with N_s oscillators. We first rewrite the Hamiltonian using matrices:

$$H = \frac{1}{2}p_j\delta_{jj'}p_{j'} + \frac{1}{2}q_jW_{jj'}q_{j'}, \tag{3.23}$$

with the symmetric matrix

$$W_{jj'} = (B + 2A)\delta_{jj'} - A(\delta_{j+1,j'} + \delta_{j-1,j'}), \tag{3.24}$$

where $\delta_{j+1,j'}$ and $\delta_{j-1,j'}$ should be understood modulo N_s (so $\delta_{N_s+1,1} = 1$). For an arbitrary symmetric matrix for the potential energy $V = \frac{1}{2}q_jW_{jj'}q_{j'}$, the Euler–Lagrange equations read

$$\ddot{q}_j = -W_{jj'}q_{j'}', \tag{3.25}$$

and it is clear that this linear problem can be solved by diagonalizing $W_{jj'}$. In the particular case (3.24) considered here, this problem can be solved by symmetry arguments. The reason we introduced periodic boundary conditions is that they guarantee that the Lagrangian is invariant under the discrete translations

$$q_j \to T_{jj'}q_{j'} = q_{j+1},$$ (3.26)

with the shift operator

$$T_{jj'} = \delta_{j+1,j'}.$$ (3.27)

The matrix T can be visualized as having 1's on the upper diagonal and the lower left corner.

$$T = \begin{pmatrix} 0 & 1 & 0 & \cdots & 0 \\ 0 & 0 & 1 & \cdots & 0 \\ \vdots & \vdots & \ddots & \ddots & \vdots \\ 0 & 0 & 0 & \cdots & 1 \\ 1 & 0 & 0 & \cdots & 0 \end{pmatrix}$$ (3.28)

This is a cyclic matrix which satisfies the relations $T^T = T^{-1}$ and $T^{N_s} = 1$. Consequently T is a orthogonal (and consequently unitary) matrix and its N_s eigenvalues are the N_s distinct N_s roots of unity

$$\lambda_l = \exp\left(i\frac{2\pi}{N_s}l\right)$$ (3.29)

with $l = 1, ...N_s$. The corresponding normalized eigenvectors are

$$v_{l,j} = \frac{1}{\sqrt{N_s}}\lambda_l^j.$$ (3.30)

Using the identity

$$(1 - y)(1 + y + \cdots + y^{N_s-1}) = 1 - y^{N_s},$$ (3.31)

we find the orthogonality relations

$$\sum_{j=0}^{N_s-1} v_{l,j}^{\star}v_{l',j} = \delta_{l,l'}.$$ (3.32)

In other words, the discrete Fourier transform is unitary and we can think of $v_{l,j}$ as a unitary matrix U_{jl}. In addition, from equations (3.24) and (3.5), we have (with no sum on l)

$$\sum_{j'=0}^{N_s-1} W_{jj'}v_{l,j'} = \omega_l^2 v_{l,j}.$$ (3.33)

Following the example of the single oscillator, we can rewrite the operators $\hat{q}_j$ and $\hat{p}_j$ as linear combinations of $\hat{A}_l$ and $\hat{A}_l^\dagger$, such that

$$\left[\hat{A}_l, \hat{A}_{l'}^\dagger\right] = \delta_{l,l'}, \text{ and } [\hat{A}_l, \hat{A}_{l'}] = \left[\hat{A}_l^\dagger, \hat{A}_{l'}^\dagger\right] = 0. \tag{3.34}$$

One can check that if we set

$$\hat{q}_j = \frac{1}{\sqrt{N_s}} \sum_{l=0}^{N_s-1} \frac{1}{\sqrt{2\omega_l}} \left(\hat{A}_l \exp\left(i\frac{2\pi}{N_s}lj\right) + \hat{A}_l^\dagger \exp\left(-i\frac{2\pi}{N_s}lj\right) \right) \tag{3.35}$$

$$\hat{p}_j = \frac{-i}{\sqrt{N_s}} \sum_{l=0}^{N_s-1} \sqrt{\frac{\omega_l}{2}} \left(\hat{A}_l \exp\left(i\frac{2\pi}{N_s}lj\right) - \hat{A}_l^\dagger \exp\left(-i\frac{2\pi}{N_s}lj\right) \right) \tag{3.36}$$

we recover equation (3.12).

Exercise 1: Check explicitly that this statement is correct.

In addition

$$W_{jj'}\hat{q}_{j'} = \frac{1}{\sqrt{N_s}} \sum_{l=0}^{N_s-1} \omega_l \sqrt{\frac{\omega_l}{2}} \left(\hat{A}_l \exp\left(i\frac{2\pi}{N_s}lj\right) + \hat{A}_l^\dagger \exp\left(-i\frac{2\pi}{N_s}lj\right) \right), \tag{3.37}$$

which implies that

$$\hat{H} = \frac{1}{2} \sum_{j=1}^{N_s} \left(\hat{p}_j^2 + A(\hat{q}_{j+1} - \hat{q}_j)^2 + B\hat{q}_j^2 \right), \tag{3.38}$$

$$= \sum_{l=1}^{N_s} \omega_l \left(\hat{A}_l^\dagger \hat{A}_l + \frac{1}{2} \right) \tag{3.39}$$

It is now easy to construct the spectrum of $\hat{H}$. Each of the $\hat{A}_l$ has its own Hilbert space just like $\hat{a}$ for the single harmonic oscillator. The energy eigenstates are

$$|n_1, n_2, \ldots, n_{N_s}\rangle \equiv |n_1\rangle \otimes |n_2\rangle \otimes \cdots \otimes |n_{N_s}\rangle, \tag{3.40}$$

with energy eigenvalues

$$E_{n_1,n_2,\ldots,n_{N_s}} = \sum_{l=1}^{N_s} \omega_l \left(n_l + \frac{1}{2} \right). \tag{3.41}$$

The vacuum corresponds to the case $n_l = 0$ for all l. We can also identify 'one-mode states' with a single $n_l = 1$ all the other $n_{l'}$ being zero with an energy ω_l above the vacuum.

3.2 The infinite volume and continuum limits

So far, we have connected N_s oscillators on a ring with nearest neighbor interactions but we have not discussed the physical meaning of the distance between the oscillators. We now introduce a spatial lattice spacing a_s. The physical positions of the oscillators on the lattice are

$$x = a_s j. \tag{3.42}$$

This does not affect the Fourier modes $\exp\left(i\frac{2\pi}{N_s}lj\right)$ provided that we also introduce physical momenta

$$k = \frac{2\pi}{N_s a_s}l. \tag{3.43}$$

Note that in PS, the symbol p is often used for the momenta, however, in this section we already use this symbol for the Hamiltonian conjugate momenta p_j. With these notations,

$$\exp\left(i\frac{2\pi}{N_s}lj\right) = \exp(ikx). \tag{3.44}$$

The physical size of the system is

$$N_s a_s = L. \tag{3.45}$$

The spacing between the allowed momenta compatible with the periodic boundary conditions is

$$\Delta k = \frac{2\pi}{N_s a_s}. \tag{3.46}$$

We first consider the infinite volume limit, keeping the lattice spacing at a finite nonzero value. We can replace the sum over the momenta by an integral:

$$\sum_l = \frac{1}{\Delta k}\sum_l \Delta k \rightarrow L\int\frac{dk}{2\pi}. \tag{3.47}$$

The range of integration should cover an interval of length $N_s\Delta k = 2\pi/a$ but there is some arbitrariness in choosing the center of the interval. A plane wave $\exp(ikx)$ has the lattice periodicity provided that k is an integer multiple of $2\pi/a_s$. This set of momenta is called the *reciprocal lattice*. In crystallography, the condition for constructive interference is that the difference between the incident and reflected x-ray momentum belongs to the reciprocal lattice. In our lattice construction, shifting the momenta by any element of the reciprocal lattice has no effect. We can take the interval of length $N_s\Delta k = 2\pi/a_s$ anywhere we want. The reciprocal lattice introduces an equivalence relation among momenta differing by an element of the reciprocal lattice and we need to pick a 'fundamental domain' of all the inequivalent momenta. It is convenient to follow the crystallographic convention of using the *first Brillouin zone*, which are the momenta closer to the origin than to any other element of the reciprocal lattice. Here, this is just the momenta between $-\pi/a_s$ and π/a_s. This interval is invariant under a parity transformation $k \to -k$ without invoking the equivalence relation.

In summary, in the limit of large volume, we can make the replacement

$$\frac{1}{N_s a_s} \sum_{l=1}^{N_s} \to \int_{-\pi/a_s}^{\pi/a_s} \frac{dk}{2\pi}. \tag{3.48}$$

If we insert 1 in both sides we obtain $1/a_s$ consistently. Because of the finite lattice spacing the momenta are cutoff at π/a_s. This is an example of ultraviolet (UV) cutoff. Crystallographers are mostly interested in exploring the features of the lattice of scattering centers and they use light that has a wavelength of order a_s which corresponds to an energy of hc/a_s. This energy is of the order of 10 keV (x-rays) when $a_s \sim 1$ Å. On the other hand, for high-energy physicists, the UV cutoff represents the scale at which some new physics may appear and they would like to eliminate the cutoff dependence for processes involving energies much smaller than the cutoff.

If we insert the Fourier modes identified in equation (3.44) in the replacement prescription (3.48), we obtain

$$\frac{1}{N_s a_s} \sum_{l=1}^{N_s} \exp\left(i\frac{2\pi}{N_s}lj\right) = \frac{1}{a_s}\delta_{j,0} \to \int_{-\pi/a_s}^{\pi/a_s} \frac{dk}{2\pi} \exp(ikx). \tag{3.49}$$

We can now take the *continuum limit* $a_s \to 0$, where we replace

$$\sum_j a_s \to \int dx. \tag{3.50}$$

In this limit, we have

$$\frac{1}{a_s}\delta_{j,0} \to \delta(ja_s) = \delta(x), \tag{3.51}$$

and we recover the standard result

$$\int_{-\infty}^{\infty} dx \exp(ikx) = 2\pi\delta(k). \tag{3.52}$$

Looking at the discretized version of this last equation at finite volume, we find the correspondence

$$L\delta_{l,0} \to 2\pi\delta(k). \tag{3.53}$$

Some care is needed in the limit of the relation between the frequencies and momenta (3.5) because the lattice spacing appears explicitly in $\cos(k_l) = \cos(ka_s)$. One possibility is to take $A = 1/a_s^2$ and $B = m^2$. In the continuum limit, we get the standard relativistic dispersion relation

$$\omega_l^2 \to \omega(k)^2 = k^2 + m^2. \tag{3.54}$$

We are now in position to turn the discrete commutation relations (3.12) into continuous ones. We define the continuous field in the limit of infinite volume and zero lattice spacing as

$$\hat{q}_j / \sqrt{a_s} \to \hat{\phi}(j a_s) = \hat{\phi}(x), \tag{3.55}$$

and similarly

$$\hat{p}_j / \sqrt{a_s} \to \hat{\pi}(j a_s) = \hat{\pi}(x). \tag{3.56}$$

Using the integral replacement (3.47) in the discrete sums (3.35) and (3.36), we obtain

$$\hat{\phi}(x) = \int \frac{dk}{2\pi} \frac{1}{\sqrt{2\omega(k)}} (\hat{a}(k) \exp(ikx) + \hat{a}(k)^\dagger \exp(-ikx)), \tag{3.57}$$

$$\hat{\pi}(x) = -i \int \frac{dk}{2\pi} \sqrt{\frac{\omega(k)}{2}} (\hat{a}(k) \exp(ikx) - \hat{a}(k)^\dagger \exp(-ikx)), \tag{3.58}$$

with

$$\sqrt{L} \hat{A}_l \to \hat{a}\left(\frac{2\pi}{N_s a_s} l\right) = \hat{a}(k). \tag{3.59}$$

The replacement of the discrete Kronecker delta in momentum space in equation (3.53) implies that

$$[\hat{a}(k), \hat{a}^\dagger(k')] = 2\pi\delta(k - k'). \tag{3.60}$$

Using the previous algebraic manipulations for the limiting results, we obtain that

$$[\hat{\phi}(x), \hat{\pi}(y)] = i\delta(x - y). \tag{3.61}$$

All the momentum integrals are understood as going from $-\infty$ to $+\infty$. This suggests that we should subtract the constant

$$\frac{1}{2} \sum_{l=1}^{N_s} \omega_l, \tag{3.62}$$

in the energy of the finite and discrete model (3.41). After this subtraction, the Hamiltonian becomes

$$\hat{H} = \int \frac{dk}{2\pi} \omega(k) \hat{a}(k)^\dagger \hat{a}(k), \tag{3.63}$$

in the continuum limit.

Exercise 2: Extend the construction to $D - 1$ spatial dimensions. Summarize what needs to be changed.

A picture of relativistic free particles emerges. We have an empty vacuum $|\Omega\rangle$ such that

$$\hat{a}(k)|\Omega\rangle = 0, \qquad (3.64)$$

for any k. One-particle states with momentum k and energy $E_k = \omega(k) = \sqrt{k^2 + m^2}$ are obtained by acting with a creation operator on the vacuum. The PS normalization is

$$|k\rangle = \sqrt{2E_k}\, a(k)^\dagger |\Omega\rangle, \qquad (3.65)$$

with the orthogonality relation

$$\langle k'|k\rangle = 2E_k 2\pi\delta(k - k'). \qquad (3.66)$$

As we will discuss in section 3.3, this normalization is Lorentz invariant.

It is convenient to use the Heisenberg picture for free fields. We first introduce the evolution operator with respect to $t = 0$ as

$$U(t) \equiv \exp(-i\hat{H}t), \qquad (3.67)$$

and define

$$\hat{\phi}(x, t) = U(t)^\dagger \hat{\phi}(x) U(t), \qquad (3.68)$$

and similarly

$$\hat{a}(k, t) = U(t)^\dagger \hat{a}(k) U(t). \qquad (3.69)$$

Since

$$\frac{d}{dt}\hat{\phi}(x, t) = i[\hat{H}, \hat{\phi}(x, t)], \qquad (3.70)$$

we only need to solve

$$\frac{d}{dt}\hat{a}(k, t) = i[\hat{H}, \hat{a}(k, t)] = -i\omega(k)\hat{a}(k, t). \qquad (3.71)$$

The solution is

$$\hat{a}(k, t) = \exp(-i\omega(k)t)\hat{a}(k), \qquad (3.72)$$

in analogy with the classical formula (3.16) and consequently

$$\hat{\phi}(x, t) = \int \frac{dk}{2\pi} \frac{1}{\sqrt{2\omega(k)}}(\hat{a}(k) \exp(-i(\omega(k)t - kx)) + \text{h.c.}), \qquad (3.73)$$

$$\hat{\pi}(x, t) = -i \int \frac{dk}{2\pi} \sqrt{\frac{\omega(k)}{2}} (\hat{a}(k) \exp(-i(\omega(k)t - kx)) - \text{h.c.}). \qquad (3.74)$$

Notice that $\frac{\partial}{\partial t}\hat{\phi}(x, t) = \hat{\pi}(x, t)$ as expected.

3.3 Free KG and Dirac quantum fields in 3 + 1 dimensions

The continuous results obtained in the previous section can be continued to an arbitrary number of spatial dimensions. This is done in three spatial dimensions in PS. We follow their notations. The momenta are denoted by $\mathbf{p}$ instead of k and $\omega(k)$ is replaced by

$$E_{\mathbf{p}} \equiv \sqrt{\mathbf{p} \cdot \mathbf{p} + m^2}. \qquad (3.75)$$

In the formulas below, the mass shell condition is imposed on the energy:

$$p \cdot x \equiv E_{\mathbf{p}}t - \mathbf{k} \cdot \mathbf{x}. \qquad (3.76)$$

With these notations,

$$\hat{\phi}(\mathbf{x}, t) = \int \frac{dp^3}{(2\pi)^3} \frac{1}{\sqrt{2E_{\mathbf{p}}}} (\hat{a}(\mathbf{p}) \exp(-ip \cdot x) + \text{h.c.}), \qquad (3.77)$$

$$\hat{\pi}(\mathbf{x}, t) = \frac{\partial}{\partial t}\hat{\phi}(\mathbf{x}, t). \qquad (3.78)$$

The commutation relations become

$$[\hat{a}(\mathbf{k}), \hat{a}^\dagger(\mathbf{k}')] = (2\pi)^3 \delta^3(\mathbf{k} - \mathbf{k}'), \qquad (3.79)$$

while the other two commutators are zero, and

$$[\hat{\phi}(\mathbf{x}), \hat{\pi}(\mathbf{y})] = i\delta^3(\mathbf{x} - \mathbf{y}). \qquad (3.80)$$

The Hamiltonian becomes

$$\hat{H} = \int \frac{d^3p}{(2\pi)^3} E_{\mathbf{p}}\hat{a}(\mathbf{p})^\dagger\hat{a}(\mathbf{p}), \qquad (3.81)$$

The one-particle states are defined as

$$|\mathbf{p}\rangle = \sqrt{2E_{\mathbf{p}}} a(\mathbf{p})^\dagger|0\rangle, \qquad (3.82)$$

with the orthogonality relation

$$\langle\mathbf{p}|\mathbf{p}'\rangle = 2E_{\mathbf{p}}(2\pi)^3\delta^3(\mathbf{p} - \mathbf{p}'). \qquad (3.83)$$

The normalization is Lorentz invariant because

$$\int \frac{d^3p}{2E_{\mathbf{p}}} = \int d^4p\, \delta(p_0^2 - \mathbf{p} \cdot \mathbf{p})\theta(p_0), \qquad (3.84)$$

is Lorentz invariant. The matrix element of the field between a one particle state and the vacuum is

$$\langle 0|\hat{\phi}(\mathbf{x}, t)|\mathbf{p}\rangle = \exp(-ip \cdot x), \qquad (3.85)$$

which can be interpreted as a wavefunction.

From equation (3.77), the '+ h.c.' makes clear that $\hat{\phi}(\mathbf{x}, t)$ is a Hermitian operator corresponding to a real classical scalar field. In section 2.4, we defined a complex scalar field $\Phi = (\phi_1 + i\phi_2)/\sqrt{2}$. Expanding the $\hat{\phi}_j$'s in terms of $\hat{a}_j$'s and defining

$$\hat{a} \equiv \frac{\hat{a}_1 + i\hat{a}_2}{\sqrt{2}} \text{ and } \hat{b} \equiv \frac{\hat{a}_1 - i\hat{a}_2}{\sqrt{2}}, \qquad (3.86)$$

we obtain

$$\hat{\Phi}(\mathbf{x}, t) = \int \frac{dp^3}{(2\pi)^3} \frac{1}{\sqrt{2E_\mathbf{p}}} (\hat{a}(\mathbf{p}) \exp(-ip \cdot x) + \hat{b}^\dagger(\mathbf{p}) \exp(+ip \cdot x)), \qquad (3.87)$$

and

$$\hat{\Phi}^\dagger(\mathbf{x}, t) = \int \frac{dp^3}{(2\pi)^3} \frac{1}{\sqrt{2E_\mathbf{p}}} (\hat{b}(\mathbf{p}) \exp(-ip \cdot x) + \hat{a}^\dagger(\mathbf{p}) \exp(+ip \cdot x)). \qquad (3.88)$$

Notice that both sets ($\hat{a}$ and $\hat{b}$) of creation and annihilation operators satisfy the standard commutation relations and commute with each other. We see that in $\hat{\Phi}(\mathbf{x}, t)$, $\hat{a}(\mathbf{p})$ comes with the positive frequency solution $\exp(-ip \cdot x)$ and $\hat{b}^\dagger(\mathbf{p})$ comes with the negative frequency solution $\exp(ip \cdot x)$. If we extend this prescription for the Dirac field, we obtain:

$$\hat{\psi}(\mathbf{x}, t) = \int \frac{dp^3}{(2\pi)^3} \frac{1}{\sqrt{2E_\mathbf{p}}}$$
$$\times \sum_{s=1,2} (\hat{a}^s(\mathbf{p})u^s(\mathbf{p}) \exp(-ip \cdot x) + \hat{b}^{s\dagger}(\mathbf{p})v^s(\mathbf{p}) \exp(+ip \cdot x)), \qquad (3.89)$$

and

$$\hat{\bar{\psi}}(\mathbf{x}, t) = \int \frac{dp^3}{(2\pi)^3} \frac{1}{\sqrt{2E_\mathbf{p}}}$$
$$\times \sum_{s=1,2} (\hat{b}^s(\mathbf{p})\bar{v}^s(\mathbf{p}) \exp(-ip \cdot x) + \hat{a}^{s\dagger}(\mathbf{p})\bar{u}^s(\mathbf{p}) \exp(+ip \cdot x)). \qquad (3.90)$$

The similarities between the expressions is due to the fact that both theories describe charged particles. It is interesting to carry the calculation of the quantum Hamiltonian in parallel. For the classical complex scalar field, we have

$$\Pi = \frac{\partial \mathcal{L}_{CKG}}{\partial \dot{\Phi}} = \dot{\Phi}^{\star}, \text{ and } \Pi^{\star} = \frac{\partial \mathcal{L}_{CKG}}{\partial \dot{\Phi}^{\star}} = \dot{\Phi}. \tag{3.91}$$

The Hamiltonian density reads

$$\mathcal{H}_{CKG} = \dot{\Phi}^{\star}\dot{\Phi} + \nabla\Phi^{\star} \cdot \nabla\Phi + m^2\Phi^{\star}\Phi. \tag{3.92}$$

In this derivation we have treated Φ and $\Phi^{\star}$ as 'independent'. The same results are obtained if we work with the two real fields. For the Dirac field,

$$\frac{\partial \mathcal{L}_D}{\partial \dot{\psi}} = i\bar{\psi}\gamma^0 = i\psi^{\dagger}, \tag{3.93}$$

and we obtain

$$\mathcal{H}_D = \bar{\psi}(-i\nabla \cdot \gamma + m)\psi. \tag{3.94}$$

In these expressions, ∇ refers to the spatial derivatives with no minus sign involved.

Plugging the Fourier expansions of the quantum fields in the Hamiltonian, we obtain

$$\hat{H}_{CKG} = \int d^3x \mathcal{H}_{CKG} = \int \frac{d^3p}{(2\pi)^3} E_{\mathbf{p}}(a^{\dagger}(\mathbf{p})a(\mathbf{p}) + b(\mathbf{p})b^{\dagger}(\mathbf{p})), \tag{3.95}$$

and

$$\hat{H}_D = \int d^3x \mathcal{H}_D = \int \frac{d^3p}{(2\pi)^3} E_{\mathbf{p}} \sum_{s=1,2} (a^{s\dagger}(\mathbf{p})a^s(\mathbf{p}) - b^s(\mathbf{p})b^{s\dagger}(\mathbf{p})), \tag{3.96}$$

Exercise 3: Calculate explicitly these two Hamiltonians.

There are some potential issues with these Hamiltonians. The first one is that the vacuum energy is infinite. For $\hat{H}_{KG}$, this can be taken care of by moving $b^{\dagger}(\mathbf{p})$ to the left of $b(\mathbf{p})$ and ignoring the commutator. This amounts to redefining the energy with respect to the term generated by the commutator. This is called 'normal ordering' and denoted by enclosing the normal ordered operator between two columns:

$$:\hat{H}_{CKG}: = \int \frac{d^3p}{(2\pi)^3} E_{\mathbf{p}}(a^{\dagger}(\mathbf{p})a(\mathbf{p}) + b^{\dagger}(\mathbf{p})b(\mathbf{p})). \tag{3.97}$$

This procedure is more problematic for $\hat{H}_D$, because if we apply a similar reasoning, the number operator for the $b(\mathbf{p})$ comes with a negative sign! This means that by adding the corresponding states we lower the energy without lower bound. This problem can be solved by assuming *anticommutation* relations:

$$\{\hat{a}^s(\mathbf{k}), \hat{a}^{r\dagger}(\mathbf{k}')\} \equiv \hat{a}^s(\mathbf{k})\hat{a}^{r\dagger}(\mathbf{k}') + \hat{a}^{r\dagger}(\mathbf{k}')\hat{a}^s(\mathbf{k}) = \delta^{rs}(2\pi)^3\delta^3(\mathbf{k} - \mathbf{k}'), \tag{3.98}$$

and

$$\{\hat{b}^s(\mathbf{k}), \hat{b}^{r\dagger}(\mathbf{k}')\} = \delta^{rs}(2\pi)^3\delta^3(\mathbf{k} - \mathbf{k}'), \tag{3.99}$$

while all the other anticommutators are zero. With this new prescription, we obtain the well-behaved normal-ordered Hamiltonian:

$$:\hat{H}_D: = \int \frac{d^3p}{(2\pi)^3} E_\mathbf{p} \sum_{s=1,2} (a^{s\dagger}(\mathbf{p})a^s(\mathbf{p}) + b^{s\dagger}(\mathbf{p})b^s(\mathbf{p})). \tag{3.100}$$

The two theories have a continuous global $U(1)$ symmetry which consist in multiplying the field ψ or Φ by $\exp i\alpha$.

Exercise 4: Derive the transformation properties of the fermion bilinears under the discrete symmetries C (charge conjugation), P (parity) and T (time reversal).

Solution. This is done in complete detail in section 3.6 of PS and summarized in a table on p 71 of their book.

3.4 The Hamiltonian formalism for Maxwell's gauge fields

The implementation of the Hamiltonian formalism for abelian gauge fields $A_\mu(x)$ is non trivial. A central issue is that the time derivative of A_0 does not appear in $\mathcal{L}^{\text{Maxwell}}$. Consequently, the canonically conjugate momentum is zero. One could argue that A_0 could simply be eliminated by a gauge transformation, however, the variation of A_0 is needed to obtain the equation of motion (in the absence of sources):

$$\partial_\mu F^{\mu 0} = 0, \tag{3.101}$$

which, in any dimensions, is equivalent to Gauss's law

$$\nabla \cdot \mathbf{E} = 0. \tag{3.102}$$

As discussed in section 2.6, it is possible to introduce a gauge condition and a symmetry breaking term in the Lagrangian density:

$$\mathcal{L}^{MGF} = -(1/4)F_{\mu\nu}F^{\mu\nu} - \frac{\lambda}{2}(\partial_\mu A^\mu)^2, \tag{3.103}$$

bearing in mind that physical results like cross sections should not depend on λ. The second term breaks the gauge symmetry and introduces terms including the time derivative of A_0.

Before calculating the conjugate momenta of $\mathcal{L}^{MGF}$, let us recall our sign conventions in natural units:

$$A^\mu: (\phi, +\mathbf{A}), \tag{3.104}$$

$$\partial_\mu: (\partial/\partial t, +\nabla), \tag{3.105}$$

$$F^{i0} = E^i, \tag{3.106}$$

$$F^{ij} = -\epsilon^{ijk}B^k \ (D = 4), \tag{3.107}$$

$$F^{12} = -B \ (D = 3). \tag{3.108}$$

These are consistent with the standard $D - 1$-vector form:

$$\mathbf{E} = -\nabla\phi - \dot{\mathbf{A}}, \tag{3.109}$$

for any D, and the dimension-dependent relations

$$\nabla \times \mathbf{A} = \begin{cases} \mathbf{B}(D = 4) \\ B(D = 3). \end{cases} \tag{3.110}$$

We can now calculate the conjugate momenta.

$$\frac{\partial\mathcal{L}^{MGF}}{\partial\dot{\mathbf{A}}} = -\mathbf{E} \equiv \boldsymbol{\pi}, \tag{3.111}$$

and

$$\frac{\partial\mathcal{L}^{MGF}}{\partial\dot{A}^0} = -\lambda\partial_\mu A^\mu \equiv \pi^0. \tag{3.112}$$

We can eliminate the time derivatives using the conjugate momenta:

$$\dot{\mathbf{A}} = \boldsymbol{\pi} - \nabla A^0, \tag{3.113}$$

$$\dot{A}^0 = -\frac{\pi^0}{\lambda} - \nabla \cdot \mathbf{A}. \tag{3.114}$$

The corresponding Hamiltonian density reads

$$\mathcal{H}^{MGF} = \boldsymbol{\pi} \cdot \dot{\mathbf{A}} + \pi^0\dot{A}^0 - \mathcal{L}^{MGF}. \tag{3.115}$$

After eliminating the time derivatives, we obtain for $D = 4$

$$\mathcal{H}^{MGF} = \frac{1}{2}\boldsymbol{\pi} \cdot \boldsymbol{\pi} + \frac{1}{2}\mathbf{B} \cdot \mathbf{B} - \boldsymbol{\pi} \cdot \nabla A^0 - \frac{1}{2\lambda}(\pi^0)^2 - \pi^0\nabla \cdot \mathbf{A}. \tag{3.116}$$

For $D = 3$, we just replace $\mathbf{B} \cdot \mathbf{B}$ by B^2. For $D = 2$, there is no magnetic field. There are four Hamilton equations. The first one has the same form for any D and reads

$$\dot{\mathbf{A}} = \frac{\partial\mathcal{H}^{MGF}}{\partial\boldsymbol{\pi}} = \boldsymbol{\pi} - \nabla A^0. \tag{3.117}$$

This is just the definition of the momenta $\boldsymbol{\pi}$ in terms of the velocities $\dot{\mathbf{A}}$. The second one is dimension dependent. For $D = 4$, we have

$$\dot{\boldsymbol{\pi}} = -\frac{\partial \mathcal{H}^{MGF}}{\partial \mathbf{A}} = -\nabla \times \mathbf{B} - \nabla \pi^0. \tag{3.118}$$

This is the Maxwell equation involving the current $\mathbf{j}$ which is set to zero with an extra term. For $D = 3$, $\mathbf{B}$ is replaced by B. To be completely explicit, for $D = 2$ this Maxwell equation reads

$$\dot{E}_x = 0. \tag{3.119}$$

For $D = 3$, we have

$$\dot{E}_x = \frac{\partial}{\partial y} B$$
$$\dot{E}_y = -\frac{\partial}{\partial x} B, \tag{3.120}$$

while for $D = 4$, we have the usual form

$$\dot{\mathbf{E}} = \nabla \times \mathbf{B}. \tag{3.121}$$

For the zeroth-component, we have Hamilton's equations

$$\dot{A}^0 = \frac{\partial \mathcal{H}^{MGF}}{\partial \pi^0} = -\frac{\pi^0}{\lambda} - \nabla \cdot \mathbf{A}, \tag{3.122}$$

which again is the known relation between velocities and momenta. Finally,

$$\dot{\pi}^0 = -\frac{\partial \mathcal{H}^{MGF}}{\partial A^0} = -\nabla \cdot \boldsymbol{\pi} = \nabla \cdot \mathbf{E}. \tag{3.123}$$

This is Gauss's law with an extra term. We can recover the standard Maxwell equations with sources set to zero if we impose $\pi^0 = 0$ or equivalently the Lorenz condition $\partial_\mu A^\mu = 0$. In view of the difficulties we encountered to get the correct physics using the Hamiltonian formalism, it is clear that the quantization of the classical model using a Hilbert space of particles carrying momenta and polarization is non-trivial.

IOP Publishing

Quantum Field Theory
A quantum computation approach
Yannick Meurice

Chapter 4

A practical introduction to perturbative quantization

4.1 Overview

So far, we have been able to quantize classical free field theories. The Lagrangian density of free field theories is quadratic in the fields and the equations of motion are linear. By using Fourier decomposition, the free fields can be re-expressed in terms of harmonic oscillators. If we use a spatial lattice approximation in a finite box, there are only a finite numbers of oscillators. In this chapter, we will introduce methods that allow us to deal with field interactions. Using symbols defined in the previous chapters, the most common examples of field interactions are $\lambda\phi^4$ ('phi-four'), $g\bar{\psi}\psi\phi$ (Yukawa), $ie\bar{\psi}\gamma^\mu\psi A_\mu$ (electromagnetic) or $\text{Tr}([A_\mu, A_\nu][A^\mu, A^\nu])$ (Yang–Mills).

In a generic way, the interactions introduce anharmonic terms coupling the independent harmonic oscillators used for free field. The main idea of this chapter is to treat these anharmonic terms as perturbations. In the case of a single harmonic oscillator, a perturbation of the form λx^4 is a textbook example for time-independent, non-degenerate perturbation theory. A classic paper on this subject is 'Anharmonic oscillator' by C M Bender and T T Wu [1] where the connection between the operator formalism and Feynman diagrams is emphasized. In field theory, Feynman diagrams provide a visually appealing way to enumerate the successive terms of the perturbative series.

Perturbation theory has unsatisfactory aspects: typically, as you increase the order in perturbation, the range of validity for the perturbation shrinks and the series has a zero radius of convergence. This behavior can be traced [2] to the large field contributions in the path-integral formulation and will be discussed in more detail in section 12.5. Another serious issue appearing in perturbative field theory, is that in the continuum limit, there is an infinite number of oscillators and typically, infinite corrections appear at the lowest order of perturbation. Physically, these infinities come from short-distance ultra-violet (UV) or large-distance infrared (IR) contributions. The UV infinities can be regularized by introducing a space–time lattice [3] or changing 'slightly' the dimensionality of space–time IR cutoff [4, 5]. The IR divergences can be handled by using a finite box with specific boundary conditions.

It is remarkable that despite these apparent issues, physicists were able to use perturbation theory to make calculations which, in a large majority of cases agree very well with experiments. Among these successes, the agreement between theory and experiment for the anomalous magnetic moment of the electron [6] and the muon (see http://pdg.lbl.gov/2019/reviews/rpp2018-rev-g-2-muon-anom-mag-moment.pdf and https://muon-g-2.fnal.gov) are expressed in parts per billion and can be considered as one of the 'jewels' of theoretical physics. Perturbation theory played a major role in the development of the standard model and the verified predictions of the existence of W and Z boson, top quark and the Brout–Englert–Higgs boson represent one of the most important accomplishments of the 20th century.

Despite the success of perturbation theory, there are problems that cannot be handled with the method. One of them is the confinement of quarks and gluons inside pions and nucleons using quantum chromodynamics (QCD). A satisfactory treatment of QCD can be achieved using lattice gauge theory [3] together with numerical methods based on importance sampling (Monte Carlo simulations). These simulations provide accurate answers for static properties of hadrons, such as their masses or form factors but are not able to handle real-time evolution. This is why we need to develop methods that are suitable for programming with quantum computers.

In the following, we emphasize more the practical aspects of the main results, namely the fact that the perturbative methods can be summarized by 'Feynman rules' that can be used to calculate cross sections and decay rates. Most of the derivations skipped in the rest of this chapter can be found in PS.

4.2 Dyson's chronological series

The crucial ingredients of perturbation theory are the interaction picture and Dyson's series. It is well-known that in any quantum mechanical problem described by a Hamiltonian $\hat{H}$, acting on a Hilbert space where the state of the system under consideration is given as $|\psi\rangle$ at $t = 0$, the time-evolution of an observable $\hat{A}$:

$$\langle A(t)\rangle \equiv \langle\psi|e^{+it\hat{H}}\hat{A}e^{-it\hat{H}}|\psi\rangle, \tag{4.1}$$

can either be expressed in the standard Schrödinger representation with

$$|\psi(t)\rangle = e^{-it\hat{H}}|\psi\rangle, \tag{4.2}$$

and written as

$$\langle A(t)\rangle = \langle\psi(t)|\hat{A}|\psi(t)\rangle, \tag{4.3}$$

or in the Heisenberg picture

$$\langle A(t) \rangle = \langle \psi | \hat{A}_H(t) | \psi \rangle, \tag{4.4}$$

where where the operators perform the time evolution:

$$\hat{A}_H(t) = e^{+it\hat{H}} \hat{A} e^{-it\hat{H}}. \tag{4.5}$$

When

$$H = H_0 + H_1, \tag{4.6}$$

with H_0 representing a solvable problem and H_1 a perturbation, we can also write

$$\langle A(t) \rangle = \langle \psi | e^{+it\hat{H}} e^{-it\hat{H}_0} \hat{A}_I(t) e^{+it\hat{H}_0} e^{-it\hat{H}} | \psi \rangle, \tag{4.7}$$

with the operators evolving with H_0 only:

$$\hat{A}_I(t) = e^{+it\hat{H}_0} \hat{A} e^{-it\hat{H}_0}. \tag{4.8}$$

This is the interaction picture which reduces to the Heisenberg picture in the case $H_1 = 0$.

In the interaction picture, the states prepared at a time t_0 evolve with the operator

$$U(t, t_0) = e^{+i(t-t_0)\hat{H}_0} e^{-i(t-t_0)\hat{H}}. \tag{4.9}$$

Taking the time derivative

$$\begin{aligned}
\frac{d}{dt} U(t, t_0) &= -ie^{+i(t-t_0)\hat{H}_0}(-\hat{H}_0 + \hat{H})e^{-i(t-t_0)\hat{H}} \\
&= -ie^{+i(t-t_0)\hat{H}_0}\hat{H}_1 e^{-i(t-t_0)\hat{H}_0} U(t, t_0) \\
&= -i\hat{H}_I(t) U(t, t_0),
\end{aligned} \tag{4.10}$$

with

$$\hat{H}_I(t) \equiv e^{+i(t-t_0)\hat{H}_0}\hat{H}_1 e^{-i(t-t_0)\hat{H}_0}. \tag{4.11}$$

We call $\hat{H}_I(t)$ the 'interaction Hamiltonian' because it is the interaction part of $\hat{H}$, namely $\hat{H}_1$, in the interaction picture. Since $\hat{H}_I(t)$ depends explicitly on time, we cannot express $U(t, t_0)$ as the usual exponential of an operator because in general for $t_1 \neq t_2$, we have $[\hat{H}_I(t_1), \hat{H}_I(t_2)] \neq 0$. This apparent ordering ambiguity can be lifted by using a chronological time-ordering prescription denoted T invented by Dyson:

$$\begin{aligned}
U(t, t_0) &= T\left(e^{-i\int_{t_0}^{t} dt' H_I(t')} \right) \\
&\equiv 1 - i\int_{t_0}^{t} dt_1 \hat{H}_I(t_1) - \int_{t_0}^{t} dt_2 \hat{H}_I(t_2) \int_{t_0}^{t_2} dt_1 \hat{H}_I(t_1) + \cdots.
\end{aligned} \tag{4.12}$$

In each of the integrals, the $\hat{H}_I(t_i)$ are ordered in increasing time when read *from right to left*. This means that at order n, the interaction Hamiltonian at the latest time t_n

always appears on the foremost left. Consequently, when we take a time derivative, we pull out a factor $\hat{H}_I(t)$ without affecting the rest of the chronological order

$$\frac{d}{dt}\int_{t_0}^{t}dt_n\hat{H}_I(t_n)\int_{t_0}^{t_{n-1}}dt_{n-1}\hat{H}_I(t_{n-1})\cdots = \hat{H}_I(t)\int_{t_0}^{t}dt_{n-1}\hat{H}_I(t_{n-1})\cdots \qquad (4.13)$$

This implies that the T-exponential provides a solution of equation (4.10). Notice the absence of $\frac{1}{n!}$ in equation (4.10). For a time-independent Hamiltonian, these factors reappear because

$$\int_{t_0}^{t}dt_n\cdots\int_{t_0}^{t_3}dt_2\int_{t_0}^{t_2}dt_1 = \int_{t_0}^{t}dt_n\cdots\int_{t_0}^{t_3}dt_2(t_2-t_0) = \cdots = \frac{(t-t_0)^n}{n!}. \qquad (4.14)$$

The time-ordered exponential can be rewritten in a more symmetric way where the $\frac{1}{n!}$ reappears:

$$T\left(e^{-i\int_{t_0}^{t}dt'\,H_I(t')}\right) = \sum_{n=0}^{\infty}\frac{(-i)^n}{n!}\int_{t_0}^{t}dt_n\cdots$$
$$\times\int_{t_0}^{t}dt_2\int_{t_0}^{t}dt_1 T(\hat{H}_I(t_n)\cdots\hat{H}_I(t_2)\hat{H}_I(t_1)). \qquad (4.15)$$

It is also possible to insert other operators in the time-ordered expression. This will be the basic method to derive perturbative expansions in field theory.

4.3 Feynman propagators, Wick's theorem and Feynman rules

The idea of using the chronological series in field theory has been very successful. In the rest of this chapter we offer a quick summary of what can be found, with details in PS.

One of the basic ingredients is the time-ordered product of two free fields in the free vacuum, also called Feynman propagator. For instance, for a real scalar field in $D = 4$

$$D_F(x-y) \equiv \langle 0|T(\phi_I(x)\phi_I(y))|0\rangle = \int\frac{d^4x}{(2\pi)^4}\frac{i}{p^2-m^2+i\epsilon}e^{-ip\cdot(x-y)}. \qquad (4.16)$$

Exercise 1: Derive the integral expression of the Feynman propagator (4.16) by integrating over p_0 and using the residue theorem to pick the poles corresponding to the Feynman prescription $i\epsilon$ in the denominator.

Solution. PS pp 29–31. PS also has a discussion of the propagator for the Dirac field on p 63.

The Feynman propagator is all we need to calculate chronological T-products of fields. Wick's theorem states that the T-product of an even number of identical fields in the vacuum can be expressed as the sum of products of all the possible propagators with all plus signs for bosonic fields and a minus sign for odd permutations in the case of fermionic fields.

Exercise 2: Prove Wick's theorem for KG and Dirac fields.

Solution. section 4.3 in PS.

As an example, for four real scalar fields, we have

$$\langle 0|T(\phi_I(x_1)\phi_I(x_2)\phi_I(x_3)\phi_I(x_4))|0\rangle = D_F(x_1 - x_2)D_F(x_3 - x_4)$$
$$+ D_F(x_1 - x_3)D_F(x_2 - x_4) \qquad (4.17)$$
$$+ D_F(x_1 - x_4)D_F(x_2 - x_3).$$

This result can be visualized by drawing four points and joining pairs with lines in the three possible ways. For the corresponding interacting theory, we have

$$\int dt H_I = \lambda \int d^D x \phi_I^4(x). \qquad (4.18)$$

If we slightly separate the four space–time points and insert the T-product of this expression in the vacuum, use Wick's theorem and take the limit where the four points coincide, we obtain the possibly singular expression

$$3\lambda (D_F(0))^4 \int d^D x. \qquad (4.19)$$

The factor 3 is called a combinatoric factor because it counts the ways to pick pairs. The expression $D_F(0)$ has ultraviolet (UV) singularities coming from the large momenta in equation (4.16) for $D \geqslant 2$. The last factor is the space–time volume and can be removed if we define the vacuum energy density. Graphically, the lowest order Feynman vacuum diagram is shown in figure 4.1. The combinatoric factor 3, can be obtained by dividing the number of ways to contract four fields 4! by the symmetry factor of the diagram S. Flipping the two propagators and exchanging them gives a symmetry factor $S = 2^3 = 8$ and we recover $3 = 24/S$. This can be pursued to higher orders.

Exercise 3: Draw the vacuum diagrams of the same theory at order 2 in λ. Calculate the symmetry factors and combinatoric factors.

Figure 4.1. First order vacuum diagram.

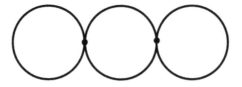

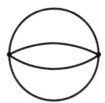

Figure 4.2. Second order vacuum diagrams.

Solution. The two diagrams are shown in figure 4.2. For the first one, $S = 2^4$ (flipping the two outside propagators, interchanging the two middle ones and exchanging the vertices) giving a combinatoric factor $(24)^2/16 = 36$. For the second one, $S = 2(4!)$ (permuting the four propagators and exchanging the vertices) giving a combinatoric factor $(24)^2/48 = 12$.

This procedure can be generalized to diagrams with external legs (fields inserted in T-products) and summarized by Feynman rules (see PS section 4.6). They are usually given in momentum space and can be summarized as follows:

    (i) For each propagator: $\frac{i}{p^2 - m^2 + i\epsilon}$;

    (ii) For each vertex: $-i24\lambda$;

    (iii) Integrate over each loop momentum: $\int \frac{d^D p}{(2\pi)^D}$;

    (iv) Divide by the symmetry factor.

Note that often authors (like PS) use $\lambda \to \lambda/24$ in order to remove the 24 in (ii) above.

4.4 Decay rates and cross sections

This is discussed extensively in PS section 4.5. Recommended exercises: derive the results in section A.3 of PS ('Gamma gymnastic'), calculate the cross section for $e^+ e^- \to \mu^+ \mu^-$ (PS section 5.1), derive all the Feynman rules in section A.1 of PS.

4.5 Radiative corrections and the renormalization program

Many loop integrals are UV divergent. This can be dealt with using a momentum cut-off or dimensional regularization (PS 7.5). These infinities can be reabsorbed in bare couplings and do not appear in physical results. Recommended exercises: work out the renormalization program for $\lambda\phi^4$ and QED at one loop level (PS chapter 10).

References

[1] Bender C M and Wu T T 1969 *Phys. Rev.* **184** 1231–60
[2] Meurice Y 2002 *Phys. Rev. Lett.* **88** 141601
[3] Wilson K 2005 *Nucl. Phys. B: Proc. Suppl.* **140** 3–19
[4] Bollini C G and Giambiagi J J 1972 *Il Nuovo Cimento B (1971–96)* **12** 20–6
[5] 't Hooft G and Veltman M 1972 *Nucl. Phys.* B **44** 189–213
[6] Aoyama T, Hayakawa M, Kinoshita T and Nio M 2015 *Phys. Rev.* D **91** 033006

IOP Publishing

Quantum Field Theory
A quantum computation approach
Yannick Meurice

Chapter 5

The path integral

5.1 Overview

The path integral method was initially developed by R Feynman [1] as a way to rewrite the quantum evolution, as a sum over paths joining some initial and final positions $\mathbf{x}_i$ and $\mathbf{x}_f$ weighted by $e^{iS/\hbar}$, where S is the action corresponding to a specific path joining $\mathbf{x}_i$ and $\mathbf{x}_f$. More specifically, the final result should be an equation of the form

$$\langle \mathbf{x}_f | e^{-i(t_f - t_i)\hat{H}/\hbar} | \mathbf{x}_i \rangle = \int [\mathcal{D}\mathbf{x}]_{\mathbf{x}_i \to \mathbf{x}_f} e^{iS[\mathbf{x}]/\hbar}, \tag{5.1}$$

where $\int [\mathcal{D}\mathbf{x}]_{\mathbf{x}_i \to \mathbf{x}_f}$ denoting the 'sum over the paths' mentioned above. In this chapter, we restore $\hbar$ in the quantum mechanical expressions. The sum over the paths can be approximated by first expressing the evolution operator as a product of N_t short evolutions by time steps a_t with $N_t a_t = (t_f - t_i)$:

$$e^{-i(t_f - t_i)\hat{H}/\hbar} = (e^{-i(a_t)\hat{H}/\hbar})^{N_t} \tag{5.2}$$

and then inserting the identity

$$\mathbb{1} = \int d\mathbf{x} |\mathbf{x}\rangle\langle\mathbf{x}|, \tag{5.3}$$

between each of the successive steps.

$$\begin{aligned}
\langle \mathbf{x}_f | e^{-i(t_f - t_i)\hat{H}/\hbar} | \mathbf{x}_i \rangle = \int d\mathbf{x}_{N_t-1} \cdots \int d\mathbf{x}_1 \langle \mathbf{x}_f | e^{-ia_t\hat{H}/\hbar} | \mathbf{x}_{N_t-1} \rangle \\
\times \langle \mathbf{x}_{N_t-1} | e^{-ia_t\hat{H}/\hbar} \cdots e^{-ia_t\hat{H}/\hbar} | \mathbf{x}_1 \rangle \\
\times \langle \mathbf{x}_1 | e^{-ia_t\hat{H}/\hbar} | \mathbf{x}_i \rangle.
\end{aligned} \tag{5.4}$$

The point of using small time steps is that, in this limit, the evolution operator matrix elements can be approximated by phases which will contribute to an overall $e^{iS/\hbar}$ factor. The product of integrals

doi:10.1088/978-0-7503-2187-7ch5

$$\int d\mathbf{x}_{N_i-1} \cdots \int d\mathbf{x}_1, \tag{5.5}$$

has a clear 'sum over the paths' interpretation illustrated in figure 5.1. However, these paths are by no mean continuous but it is somehow possible to argue that the 'important paths' are, to some extent, smooth under appropriate circumstances. This is best seen by thinking of the time parameter as a complex number and using a Euclidean time τ according to the substitution (also called Wick rotation)

$$(t_f - t_i) \rightarrow -i\tau. \tag{5.6}$$

Notice that the evolution operator is very well behaved in the time complex upper half plane when the spectrum of $\hat{H}$ is bounded from below. Obviously, the use of Euclidean time does not affect the spectrum of $\hat{H}$ and its use at large positive τ makes the contributions of the high energy states negligible. As we will see in the coming sections, the use of Euclidean time brings the substitution

$$e^{iS/\hbar} \rightarrow e^{-S/\hbar}, \tag{5.7}$$

which can be interpreted as a positive Boltzmann weight. This allows us to use importance sampling methods developed in statistical mechanics. The minima of S can be used as saddle points for the path-integral and under appropriate circumstances, one can argue that the classical solutions dominate the path integral and that the dominating fluctuations about them are, to some extent, smooth. The discretization of Euclidean time generalizes in a 'relativistic' way to field theory because space and time discretization can be treated on equivalent footing.

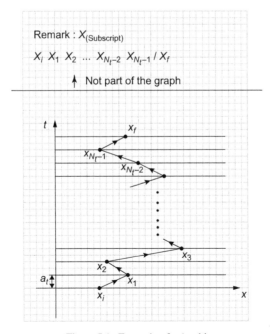

Figure 5.1. Example of a 'path'.

We will start with 'free particles' in quantum mechanics (QM) and quantum field theory (QFT). However, the reader should be warned about the fact that the (improper) states $|\mathbf{p}\rangle$ used in QM and QFT have a very different interpretation. In both cases, their use is related to a translation invariance but in QM, the translation is a shift in the argument of the wave function while in QFT, it shifts the spatial labels of the fields as explained in equation (5.49). In other words, $\mathbf{x}$ and $\mathbf{p}$ have a different meaning in the two contexts.

It is recommended to complement this chapter by reading chapter 9 in PS book.

5.2 Free particle in quantum mechanics

Our starting point for ordinary quantum mechanics with one degree of freedom will be the exact identity:

$$\langle x_n | e^{-it\hat{H}/\hbar} | x_0 \rangle = \prod_{j=1}^{n-1} \int_{-\infty}^{+\infty} dx_j \langle x_n | e^{-i\Delta t \hat{H}/\hbar} | x_{n-1} \rangle \cdots$$
$$\times \langle x_1 | e^{-i\Delta t \hat{H}/\hbar} | x_0 \rangle. \tag{5.8}$$

The integrals can be done exactly for a particle moving freely on an infinite line and described by the Hamiltonian

$$\hat{H}_0 = \frac{\hat{p}^2}{2m}. \tag{5.9}$$

Using the momentum eigenstates

$$\langle x|p\rangle = \frac{1}{\sqrt{2\pi\hbar}} e^{i\frac{p \cdot x}{\hbar}}, \tag{5.10}$$

we obtain the integral representation of the evolution operator

$$\langle x_n | \hat{U}_0(t) | x_0 \rangle = \langle x_n | e^{-it\hat{H}_0/\hbar} | x_0 \rangle = \int_{-\infty}^{+\infty} \frac{dp}{2\pi\hbar} e^{-\frac{i}{2\hbar}\left(\frac{p^2}{m}t - 2p\Delta x\right)}, \tag{5.11}$$

with $\Delta x \equiv x_n - x_0$. Completing the square in the exponential

$$\frac{p^2}{m}t - 2p\Delta x = \left(p\sqrt{\frac{t}{m}} - \Delta x \sqrt{\frac{m}{t}}\right)^2 - \Delta x^2 \frac{m}{t}, \tag{5.12}$$

and using a complex version of the standard Gaussian integration

$$\int_{-\infty}^{+\infty} du \, e^{-iBu^2} = \sqrt{\frac{\pi}{iB}}, \tag{5.13}$$

for real B, that will be justified in section 5.3, we obtain

$$\langle x_n | e^{-it\hat{H}_0/\hbar} | x_0 \rangle = \sqrt{\frac{m}{i2\pi\hbar t}} e^{i\frac{m\Delta x^2}{2\hbar t}}. \tag{5.14}$$

Using similar integration methods, we can check explicitly that

$$\langle x_n|\hat{U}_0(t_1 + t_2)|x_0\rangle = \langle x_n|\hat{U}_0(t_1)\hat{U}_0(t_2)|x_0\rangle$$
$$= \int_{-\infty}^{+\infty} dx \langle x_n|\hat{U}_0(t_1)|x\rangle\langle x|\hat{U}_0(t_2)|x_0\rangle. \tag{5.15}$$

Exercise 1: Check equation (5.15) by performing the Gaussian integration.

Solution. Complete the square in the argument of the exponential, shift the integration variable, use the formula for Gaussian integration and $\frac{1}{t_1} + \frac{1}{t_2} = \frac{t_1 + t_2}{t_1 t_2}$.

By using equation (5.15) enough time, we can verify that

$$\sqrt{\frac{m}{i2\pi\hbar t}}\, e^{i\frac{m(x_n - x_0)^2}{2\hbar t}} = \left(\sqrt{\frac{m}{i2\pi\hbar\Delta t}}\right)^n \prod_{j=1}^{n-1} \int_{-\infty}^{+\infty} dx_j e^{\frac{i}{\hbar}S_0}. \tag{5.16}$$

with

$$S_0 = \sum_{j=0}^{n-1} \Delta t \frac{m}{2}\left(\frac{x_{j+1} - x_j}{\Delta t}\right)^2. \tag{5.17}$$

Since there is no potential energy in this problem, S can be interpreted as an approximate form of the action

$$\int dt \frac{m}{2}\dot{x}^2.$$

5.3 Complex Gaussian integrals and Euclidean time

By taking the product of two Gaussian integrals with the same integrand e^{-Ax^2} with A real and positive, one can use polar coordinates and perform the integrals analytically. This yields the well-known result:

$$\int_{-\infty}^{\infty} dx e^{-Ax^2} = \sqrt{\pi/A}. \tag{5.18}$$

Since e^{-Ax^2} is entire, its integral over a close complex contour Γ is zero:

$$\oint_\Gamma dz e^{-Az^2} = 0. \tag{5.19}$$

In order to obtain a result for $A \to e^{i2\theta}A$, we use the contour shown in figure 5.2. It has four parts. The first one is the real axis between $-R$ and R, the second part is this axis rotated by an angle θ and then the two circular caps with an angle θ. The integrand in the circular caps is $e^{-AR^2 e^{i2\theta}}$ and so as long as $\cos(2\theta)$ is strictly positive, the integrand will become exponentially small when $R \to \infty$ and their integral vanish. In this limit, the contribution of the second part (the tilted axis) is

$$-\int_{-\infty}^{\infty} du e^{i\theta} e^{-Ae^{i2\theta}u^2},$$

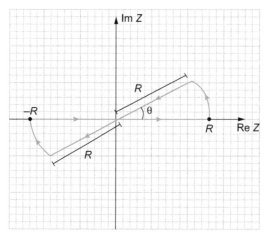

Figure 5.2. The complex contour Γ.

while the integral on the first part is just $\sqrt{\pi/A}$. Combining these results with the fact that their sum vanishes according to equation (5.19), we find the simple result

$$\int_{-\infty}^{\infty} du e^{-Ae^{i2\theta}u^2} = \sqrt{\pi/(Ae^{i2\theta})}. \tag{5.20}$$

In other words, as long as $-\frac{\pi}{4} < \theta < \frac{\pi}{4}$, $\text{Re}(Ae^{i2\theta}) > 0$, and we can just substitute $A \to Ae^{i2\theta}$ in the standard Gaussian integral formula. If we extend this result by continuity for $\theta \to \pm\frac{\pi}{4}$, we obtain

$$\int_{-\infty}^{\infty} du e^{\pm iAu^2} = \sqrt{\pi/(2A)}(1 \pm i). \tag{5.21}$$

Note that complex Gaussian integrals appear in optics where wave amplitudes corresponding to a source far from an aperture and a screen of observation not very far from the aperture (Fresnel diffraction). In this regime, wave amplitude for an aperture at a horizontal distance X from the screen and a vertical position Y on the screen is [2]

$$u(Y) = K \int_{\text{aperture}} dy e^{i\frac{2\pi}{\lambda}\frac{(Y-y)^2}{2X}}, \tag{5.22}$$

where λ is the wavelength of the incident plane waves and K a constant depending on the incident flux. This integral can be expressed in terms of the Fresnel integrals that we define with the simplest normalization

$$S(x) = \int_0^x dt \sin(t^2), \quad \text{and} \quad C(x) = \int_0^x dt \cos(t^2). \tag{5.23}$$

From the graph of $\sin(t^2)$ in figure 5.3, it is clear that the $S(x)$ gets its largest contributions from small values of t. For large x, the contributions of large t oscillate rapidly and approximately cancel.

This intuitive argument is supported by the numerical integration for $S(x)$ shown in figure 5.4. In addition, it confirms that the limit exists and in agreement with equation (5.21) we have $S(\infty) = \sqrt{\pi/8}$.

It is customary to display $S(x)$ versus $C(x)$ in a parametric plot often called the 'Cornu spiral'. We can extend the integral definition to negative values of x. We have the simple relation $C(-x) = -C(x)$ and $S(-x) = -S(x)$. In addition, equation (5.21) implies that $C(\infty) = \sqrt{\pi/8}$. Numerical results are shown in figure 5.5. The slow convergence towards the asymptotic value $(\sqrt{\pi/8}, \sqrt{\pi/8})$ is illustrated in figure 5.6. The radius of the approximate circles decreases like $1/x$.

Using these integrals we can express the light intensity for the diffraction by an edge as [2]

$$I(y) = |K'|^2[(C(-\infty) - C(y))^2 + (S(-\infty) - S(y))^2]. \tag{5.24}$$

This function is displayed in figure 5.7 and appears to be a good fit of observations [2]. This provides a physical argument that the integrals converge to limits that can be calculated using our formulas, for instance $S(\infty) = \sqrt{\pi/8}$.

Figure 5.3. $\sin(t^2)$ versus t.

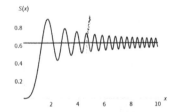

Figure 5.4. $S(x)$ versus x; the horizontal line is $\sqrt{\pi/8}$.

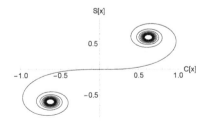

Figure 5.5. Cornu spiral: $S(x)$ versus $C(x)$ for $-10 \leqslant x \leqslant 10$.

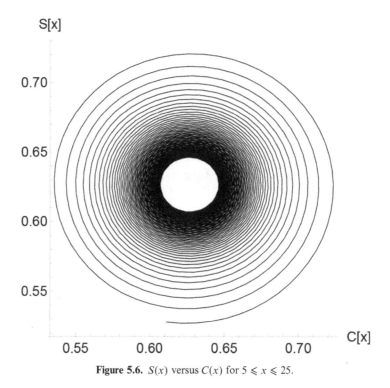

Figure 5.6. $S(x)$ versus $C(x)$ for $5 \leqslant x \leqslant 25$.

Figure 5.7. Diffraction pattern for an edge following equation (5.24). This picture was made using Mathematica, but it is a good rendering of what can be seen in a classroom by shining a laser beam on a sharp edge.

***Exercise* 2**: Reproduce figures 5.3–5.7. This can be done with the following Mathematica notebook.

In[1]:= **(∗Fig. 5∗)**
    Plot[Sin[t^2], {t, 0, 10}, PlotStyle → {RGBColor[0, 0, 0], Thickness[0.006]},
    AxesLabel → {t, Sin[t^2]}, LabelStyle → {"Text", RGBColor[0, 0, 0]}]

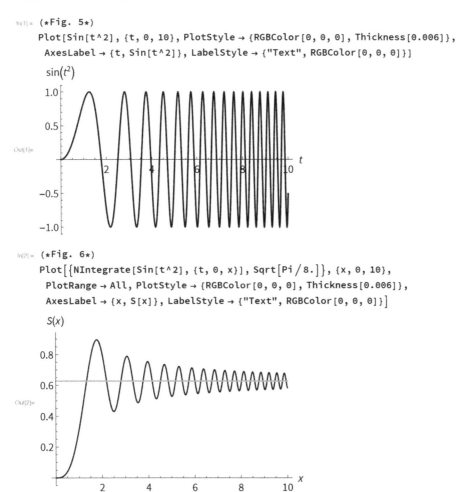

In[2]:= **(∗Fig. 6∗)**
    Plot[{NIntegrate[Sin[t^2], {t, 0, x}], Sqrt[Pi / 8.]}, {x, 0, 10},
    PlotRange → All, PlotStyle → {RGBColor[0, 0, 0], Thickness[0.006]},
    AxesLabel → {x, S[x]}, LabelStyle → {"Text", RGBColor[0, 0, 0]}]

```
In[3]:= lim = Sqrt[Pi / 8.];
       betterfr[x_, y_] := (NIntegrate[Cos[t^2], {t, 0, x}] + lim)^2 +
           (NIntegrate[Sin[t^2], {t, 0, x}] + lim)^2;
       (*Plotting for y=0*)
       Plot[betterfr[x, 0], {x, -10, 10}]
```

Out[5]=

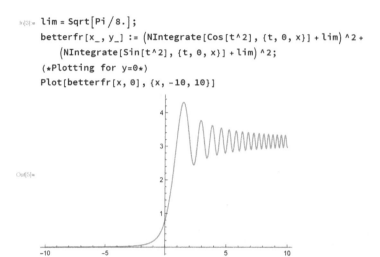

```
       (*Fig. 9; use PlotPoints→100 for a nice uniform picture,
       but it takes a while to build*)
       DensityPlot[betterfr[x, y], {x, -10, 10}, {y, 0, 0.1}, PlotRange → All,
        ColorFunction → (Hue[1, 0.8, #] &), PlotPoints → 100, Axes → False, Frame → False]
```

Out[6]=

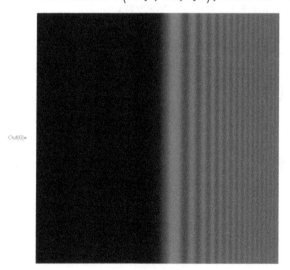

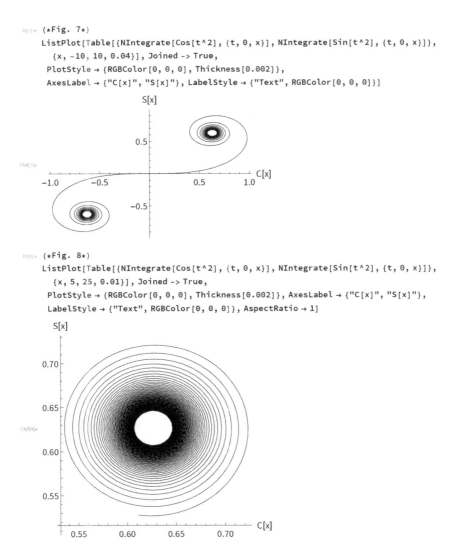

```
ln[7]= (*Fig. 7*)
    ListPlot[Table[{NIntegrate[Cos[t^2], {t, 0, x}], NIntegrate[Sin[t^2], {t, 0, x}]},
      {x, -10, 10, 0.04}], Joined -> True,
    PlotStyle -> {RGBColor[0, 0, 0], Thickness[0.002]},
    AxesLabel -> {"C[x]", "S[x]"}, LabelStyle -> {"Text", RGBColor[0, 0, 0]}]
```

```
ln[8]= (*Fig. 8*)
    ListPlot[Table[{NIntegrate[Cos[t^2], {t, 0, x}], NIntegrate[Sin[t^2], {t, 0, x}]},
      {x, 5, 25, 0.01}], Joined -> True,
    PlotStyle -> {RGBColor[0, 0, 0], Thickness[0.002]}, AxesLabel -> {"C[x]", "S[x]"},
    LabelStyle -> {"Text", RGBColor[0, 0, 0]}, AspectRatio -> 1]
```

5.4 The Trotter product formula

We now consider the quantum mechanical case of a particle in a potential:

$$\hat{H} = \hat{H}_0 + \hat{V}. \tag{5.25}$$

$\hat{V}$ is diagonal in the $|\mathbf{x}\rangle$ representation which we have used extensively but it does not commute with H_0 and

$$e^{-\frac{i}{\hbar}(\hat{H}_0 + \hat{V})t} \neq e^{-\frac{i}{\hbar}\hat{H}_0 t}e^{-\frac{i}{\hbar}\hat{V}t}. \tag{5.26}$$

However for a short amount of time ϵ,

$$e^{-\frac{i}{\hbar}(\hat{H}_0+\hat{V})\epsilon} \simeq e^{-\frac{i}{\hbar}\hat{H}_0\epsilon}e^{-\frac{i}{\hbar}\hat{V}\epsilon} + \mathcal{O}(\epsilon^2). \tag{5.27}$$

The corrections of order ϵ^2 can be calculated

$$e^{-\frac{i}{\hbar}(\hat{H}_0+\hat{V})\epsilon} \simeq e^{-\frac{i}{\hbar}\hat{H}_0\epsilon}e^{-\frac{i}{\hbar}\hat{V}\epsilon}e^{\frac{\epsilon^2}{2\hbar^2}[\hat{H}_0,\,\hat{V}]} + \mathcal{O}(\epsilon^3). \tag{5.28}$$

Exercise 3: Derive equation (5.28).

Solution. Expand the exponentials up to order ϵ^2 keeping all the products in the correct order. Expand each side up to order ϵ^2.

Equation (5.28) shows more specifically how the commutator of the kinetic and potential energy prevents us from factorizing the exponentials as we do for exponentials of c-numbers (complex numbers as opposed to operators). It can be used to construct 'improved actions' that reduce the discretization error. However, if we take the limit of a large number of arbitrarily small steps, this will not be needed.

We are now in position to state Trotter's product formula: if A and B are two $M \times M$ matrices,

$$\left(e^{\frac{A+B}{n}}\right)^n \xrightarrow[n\to\infty]{} \left(e^{\frac{A}{n}}e^{\frac{B}{n}}\right)^n. \tag{5.29}$$

In other words, in the limit where n, the number of steps of size $\frac{1}{n}$, becomes large, we can actually factorize the exponentials. As we will proceed to explain, the convergence is in matrix norm. The proof of this theorem for finite matrices and its extension to operators on Hilbert spaces is given in the Reed and Simon textbook [3]. The idea of the proof is to use the matrix identity

$$S_n^n - T_n^n = \sum_{m=0}^{n-1} S_n^m(S_n - T_n)T_n^{n-m-1}, \tag{5.30}$$

with

$$S_n = e^{\frac{A+B}{n}}, \tag{5.31}$$

and

$$T_n = e^{\frac{A}{n}}e^{\frac{B}{n}}. \tag{5.32}$$

Given a matrix norm $\| \ \|$, we can then prove that

$$\|S_n^n - T_n^n\| \leqslant n(max(\|S_n\|, \|T_n\|))^{n-1} \|S_n - T_n\|. \tag{5.33}$$

Since the error on one step is of second order, we expect

$$\|S_n - T_n\| \leqslant \frac{C}{n^2}, \tag{5.34}$$

and $\|S_n^n - T_n^n\|$ can be made as small as we want by increasing n. The constant of proportionality can be estimated and it is, in principle, possible to try to decrease the Trotter error by adding corrections using a power series expansion of

$$e^{\epsilon(A+B)} = e^{\epsilon A}e^{\epsilon B}e^{\epsilon^2 G_2}e^{\epsilon^3 G_3} \dots \tag{5.35}$$

One finds that

$$G_2 = -\frac{1}{2}[A, B]. \tag{5.36}$$

Consequently

$$\|S_n - T_n\| \simeq \frac{1}{2n^2}\|[A, B]\|, \tag{5.37}$$

when n is large enough.

For practical purposes, we need to pick a matrix norm. For any complex number a and complex matrices A and B, a matrix norm $\|\cdot\|$ should satisfy $\|aA\| = |a|\|A\|$, $\|A + B\| \leqslant \|A\| + \|B\|$ (triangle inequality), $\|A\| \geqslant 0$ and equal to zero only if $A = 0$. An example is the spectral norm of A which is the square root of the largest eigenvalue of $A^\dagger A$.

We can now extend equation (5.16) to the interacting case

$$\langle x_n|e^{-it\hat{H}/\hbar}|x_0\rangle = \left(\sqrt{\frac{m}{i2\pi\hbar\Delta t}}\right)^n \prod_{j=1}^{n-1} \int_{-\infty}^{+\infty} dx_j e^{\frac{i}{\hbar}S}. \tag{5.38}$$

with

$$S = \sum_{j=0}^{n-1} \Delta t\left[\frac{m}{2}\left(\frac{x_{j+1} - x_j}{\Delta t}\right)^2 - \frac{1}{2}(V(x_{j+1}) + V(x_j))\right], \tag{5.39}$$

where we have split the potential energy equally between the two adjacent free evolution operators. Notice that S can be seen as a discretized version of

$$S = \int dt L = \int dt\left(\frac{m}{2}\dot{x}^2 - V(x)\right). \tag{5.40}$$

We have now given a concrete meaning to equation (5.1) and we are ready to discuss specific examples and extend the results to field theory.

5.5 Models with quadratic potentials

The case of path integrals with quadratic potentials is of special interest because the discretized action (5.39) becomes a symmetric quadratic form. We can then diagonalize the corresponding matrix and obtain an exact solution, which is the focus of this subsection.

In addition, we have seen that we can safely rotate Gaussian integrals from real time to Euclidean time without losing information. Under a Wick rotation $t \to -i\tau$, we have

$$iS = i \int dt \left(\frac{m}{2} \dot{x}^2 - V(x) \right) \rightarrow - \int d\tau \left(\frac{m}{2} \left(\frac{dx}{d\tau} \right)^2 + V(x) \right) = -S_E, \qquad (5.41)$$

and the weight in the path integral becomes e^{-S_E} which is a crucial feature to use in importance sampling, as will be discussed in section 5.9.

For these reasons, we will first discuss real Gaussian integration with N real variables which will be combined in an N-dimensional vector denoted $\mathbf{X}$. We shall couple these variables to an N-dimensional vector of 'external sources' denoted $\mathbf{J}$. If $\mathcal{K}$ is a symmetric $N \times N$ matrix with strictly positive eigenvalues, one can show that

$$Z[\mathbf{J}] \equiv \int d\mathbf{X}^N e^{-\frac{1}{2}\mathbf{X}^T \mathcal{K} \mathbf{X} + \mathbf{X}^T \mathbf{J}} = \frac{(2\pi)^{N/2}}{(det\ \mathcal{K})^{1/2}} e^{+\frac{1}{2}\mathbf{J}^T \frac{1}{\mathcal{K}} \mathbf{J}}. \qquad (5.42)$$

Exercise 4: Derive equation (5.42).

Solution. As $\mathcal{K}$ is a symmetric matrix, we can diagonalize it with an orthogonal transformation A:

$$A^T \mathcal{K} A = \Lambda, \qquad (5.43)$$

with Λ a diagonal matrix with strictly positive entries. Changing variables $\mathbf{X} = A\mathbf{X}'$ and completing the square, we can perform the N Gaussian integrals independently and the desired result follows.

Exercise 5: The simple harmonic oscillator. Using the discrete action equation (5.39) for $V(x) = \frac{1}{2}m\omega^2 x^2$ and the general result (5.42), calculate $\langle x_f | e^{-it\hat{H}/\hbar} | x_i \rangle$.

Elements of the solution: x_i and x_f should be treated as sources. Use the Chebyshev's polynomials of the second kind $U_n(x)$ to determine the eigenvectors of the quadratic form. They satisfy the recursion relation

$$U_{n+1}(x) + U_{n-1}(x) = 2x U_n(x), \qquad (5.44)$$

with the initial conditions $U_0(x) = 1$ and $U_1(x) = 2x$. They can also be written as

$$U_n(x) = \frac{\sin((n+1)\arccos(x))}{\sin(\arccos(x))}. \qquad (5.45)$$

Exercise 6: Take the time continuum limit in the result for the path integral of the harmonic oscillator. Discuss energy eigenvalues and eigenfunctions.

Solution.

$$\langle x_f | e^{-\frac{H\tau}{\hbar}} | x_i \rangle = \sqrt{\frac{m\omega_0}{2\pi\hbar \sinh(\omega_0\tau)}} e^{-S_{\text{class.}}(x_i,\ x_f)}, \qquad (5.46)$$

with

$$S_{\text{class.}}(x_i,\ x_f) = \frac{m\omega_0}{2\hbar \sinh(\omega_0\tau)}((x_i^2 + x_f^2)\cosh(\omega_0\tau) - 2x_i x_f). \qquad (5.47)$$

One can insert the identity as a complete sum of eigenstates. By tracing one obtains

$$\mathrm{Tr}\, e^{-\frac{H\tau}{\hbar}} = \frac{1}{2\sinh(\omega_0\tau/2)}, \tag{5.48}$$

either by summing over the known energies or performing the Gaussian integral over $x = x_i = x_f$. Eigenfunctions can be obtained by expanding the hyperbolic functions in terms of exponentials in the untraced expression.

5.6 Generalization to field theory

We have seen before that free field theories can be envisioned as a collection of harmonic oscillators and we are now equipped to deal with these kind of problems using Gaussian integrations. To transition to field theory, we just need to adapt the scheme used in equation (5.49) to the lattice setting and replace the temporal sites j by those of a space–time lattice $x = (t, \mathbf{x})$. We will make this transition more precise in chapter 6. As before, one should avoid the possible confusion between the position operators in quantum mechanics and the position labels in field theory.

$$\hat{x} \to \hat{\phi}_{\mathbf{x}},$$

$$\sum_j \to \sum_x,$$

$$x_{j+1} - x_j \to \phi_{x+\hat{\mu}} - \phi_x.$$

For a space–time lattice with V sites and $\mathcal{K}$ a discrete Euclidean version of $\Box + m^2$, we can use (5.42) and obtain

$$Z[\mathbf{J}] \equiv \int d\phi^V e^{-\frac{1}{2}\phi^T \mathcal{K}\phi + \phi^T \mathbf{J}} = \frac{(2\pi)^{V/2}}{(det\, \mathcal{K})^{1/2}} e^{+\frac{1}{2}\mathbf{J}^T \frac{1}{\mathcal{K}}\mathbf{J}}. \tag{5.49}$$

As $Z[\mathbf{J}]$ is obtained by integrating over all the possible field configurations $\{\phi_x\}$ seen as functions of x, we often call it the 'functional integral'. We can interpret

$$\mathcal{D} \equiv \frac{1}{\mathcal{K}}, \tag{5.50}$$

as a discrete Euclidean version of the Feynman propagator (4.16).

We can also calculate the so-called free n-point functions

$$\langle \phi_{x_1} \cdots \phi_{x_n} \rangle \equiv \frac{1}{Z[0]} \int d\phi^V e^{-\frac{1}{2}\phi^T \mathcal{K}\phi} \phi_{x_1} \cdots \phi_{x_n}. \tag{5.51}$$

They can be calculated using functional derivatives with respect to the sources

$$\langle \phi_{x_1} \cdots \phi_{x_n} \rangle = \frac{1}{Z[0]} \frac{\partial}{\partial J_{x_1}} \cdots \frac{\partial}{\partial J_{x_n}} Z[\mathbf{J}]|_{\mathbf{J}=0}. \tag{5.52}$$

This leads to a functional form of Wick's theorem: assuming n even, the n-point functions are sums of products of $n/2$ propagators with the indices corresponding to all the possible ways to pick the pairs. As an example,

$$\langle \phi_{x_1}\phi_{x_2}\phi_{x_3}\phi_{x_4}\rangle = \mathcal{D}_{x_1x_2}\mathcal{D}_{x_3x_4} + \mathcal{D}_{x_1x_3}\mathcal{D}_{x_2x_4} + \mathcal{D}_{x_1x_4}\mathcal{D}_{x_2x_3}. \tag{5.53}$$

Exercise 7: Derive equation (5.53).

Solution. Using the fact that $\mathcal{K}$ and its inverse $\mathcal{D}$ are symmetric,

$$\frac{\partial}{\partial J_{x_1}}e^{+\frac{1}{2}\mathbf{J}^T\mathcal{D}\mathbf{J}} = e^{+\frac{1}{2}\mathbf{J}^T\mathcal{D}\mathbf{J}}\sum_x \mathcal{D}_{x_1x}J_x. \tag{5.54}$$

At the end, we will set the sources to zero, so J_x needs to be removed by one of the three other derivatives. Repeating with the two other derivatives we reach the announced result. The reasoning extends easily to an arbitrary even number of fields.

5.7 Functional methods for interactions and perturbation theory

The introduction of interactions in field theory is very similar to what we did in quantum mechanics. For instance for $\lambda\phi^4$ interactions, we can insert a factor

$$e^{-\lambda\sum_x a^D\phi_x^4}, \tag{5.55}$$

in the functional integral.

$$Z_{int}[\mathbf{J}, \lambda] \equiv \int d\phi^V e^{-\frac{1}{2}\phi^T\mathcal{K}\phi + \phi^T\mathbf{J}}e^{-\lambda\sum_x a^D\phi_x^4}. \tag{5.56}$$

Furthermore, it is possible to replace the field by the source derivative acting on the free theory exponential:

$$\frac{\partial}{\partial J_x}e^{-\frac{1}{2}\phi^T\mathcal{K}\phi + \phi^T\mathbf{J}} = \phi_x e^{-\frac{1}{2}\phi^T\mathcal{K}\phi + \phi^T\mathbf{J}}. \tag{5.57}$$

Using this property, we can rewrite

$$Z_{int}[\mathbf{J}, \lambda] = e^{-\lambda\sum_x a^D\left(\frac{\partial}{\partial J_x}\right)^4}Z[\mathbf{J}]. \tag{5.58}$$

Expanding in powers of λ we can rederive the Feynman rules.

It is a good idea to build up familiarity with these functional methods using simple integrals and quantum mechanics.

Exercise 8: Consider the 0-dimensional field theory with $Z \equiv \int_{-\infty}^{+\infty} d\phi e^{-\frac{1}{2}m^2\phi^2 - \lambda\phi^4}$. Calculate Z and $\ln(Z)$ up to order 3 in λ. Using functional methods, re-derive these results using Feynman's rules. Justify all the symmetry factors.

Exercise 9: Consider the anharmonic oscillator with

$$H = \frac{p^2}{2m} + \frac{1}{2}m\omega^2 x^2 + \lambda x^4. \tag{5.59}$$

Calculate the vacuum energy at first order in λ using (1) conventional perturbative methods in quantum mechanics, (2) the Feynman diagram method. Order 2 is optional.

So far we have focused on the KG field. A generalization to the Dirac and Maxwell fields will be discussed in chapter 6.

5.8 Maxwell's fields at Euclidean time

In the continuum, the transition from Minkowskian (real) time to Euclidean time for Maxwell's field is summarized in the table below.

Minkowskian	Euclidean
t	$-i\tau$
e^{-it}	$e^{-\tau}$
$g^{\mu\nu}$	$\delta^{\mu\nu}$
$d^D x$	$-i d^D x_E$
$\mathbf{x}$	$\mathbf{x}$
∇	∇
$\mathbf{A}$	$\mathbf{A}$
$\mathbf{B}$	$\mathbf{B}$
ϕ	$-iA^D$
$\dfrac{\partial \mathbf{E}_M}{\partial t}$	$-\dfrac{\partial \mathbf{E}_E}{\partial \tau}$

The last two lines of the table require some explanations. At Euclidean time, we keep the Minkowskian definition of the field strength tensor

$$F^{\mu\nu} = \partial^\mu A^\nu - \partial^\nu A^\mu,$$

but all the indices can be raised or lowered without changing the sign. We define

$$E_E^j = F_E^{jD}, \tag{5.60}$$

which should transform like $\partial/\partial t \to i\partial/\partial \tau$ under $t \to -i\tau$. This can be accomplished with $\phi \to -iA^D$ because we can lower the index of the spatial index without changing the sign as done in Minkowski space (see equation (3.109)). As we keep the standard relation $\nabla \times \mathbf{A} = \mathbf{B}$, which involves only three-vectors, we have for the same reason

$$F_E^{jk} = +\epsilon^{jkl} B_E^l. \tag{5.61}$$

With these definitions, the sign in Maxwell equation (3.121) changes.

$$\dot{\mathbf{E}}_E = -\nabla \times \mathbf{B}_E. \tag{5.62}$$

5.9 Connection to statistical mechanics

One important aspect of using Euclidean time in the path integral is that we can interpret $\exp(-S_E(\{\phi\}))$ as a probability density for the field configuration $\{\phi\}$. This probability density can be interpreted as a Boltzmann weight and allows us to use importance sampling. Consider the Ising model on a 512 by 512 lattice. Macroscopic averages are defined by summing over $2^{262\,144} \sim 10^{79\,000}$ configurations. It is clearly unfeasible to loop through such a large number of configurations with any kind of computer. However, we can obtain good approximations by picking a limited number of 'important' configurations which can capture most of the physical behavior. The power of the method has been a subject of fascination for some mathematicians such as C Villani [4].

A simple method for importance sampling can be summarized as follows. More details and references can be found in reference [5]. We just consider the Boltzmann weight for the Ising model $\sigma_x = \pm 1$ with an energy

$$E = -\sum_{x,\mu} \sigma_{x+\mu}\sigma_x, \tag{5.63}$$

and an inverse temperature β (for notations regarding sites and links of the lattice, see chapter 6). We start with a random spin configuration and update with the following procedure:

- Pick a spin at random and calculate ΔE resulting from flipping it;
- If $\Delta E < 0$, flip the spin;
- If $\Delta E > 0$, flip the spin with a probability $\exp(-\beta\Delta E)$;
- Repeat until averages become stable.

This can be implemented with a few lines of python code, as illustrated in upcoming exercises. Looking at configurations for the two-dimensional Ising model for different β provides a good insight regarding the formation of domains as shown in figure 5.8.

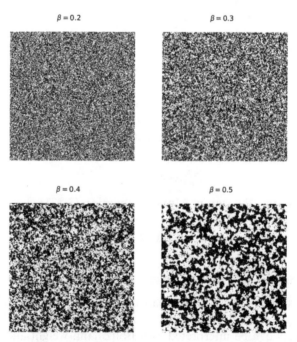

Figure 5.8. Ising configurations for increasing β (diminishing temperature).

5.10 Simple exercises on random numbers and importance sampling

Before using a random number generator (RNG) in importance sampling, it is interesting to test and compare RNGs with simple numerical experiment, such as coin tossing, where the probabilities can be calculated exactly.

Exercise 10: Using Python, perform numerical simulations corresponding to the experiment of figure 6.2 in the Feynman *Lectures of Physics vol I* (histogram of the number of heads for 100 games of 30 tosses of a fair coin; see https://www.feynmanlectures.caltech.edu/I_06.html). Display the results graphically. Are your results and figure 6.2 compatible? Is your random number generator a good substitute for a fair coin? With the existing statistics, can you tell the difference between a 0.5/0.5 coin and a 0.45/0.55 coin? What is the total number of heads for the 30 000 tosses in your numerical experiment? How far from 15 000 can it reasonably be for a fair coin? From the data, can you estimate the best probability for a head in one toss?

Solution. From basic combinatorics, if we have the probability p_h for a head in one toss, the probability for l heads in m tosses is

$$P(l \text{ heads in } m \text{ tosses}) \equiv p(l, m, p_h) = (p_h)^l (1 - p_h)^{m-l} \frac{m!}{l!(m - l)!}. \tag{5.64}$$

In order to estimate the expected number of heads in m tosses, we add up the number of heads multiplied by their probability. For m independent tosses we have

$$\sum_{l=0}^{m} l \frac{m!}{l!(m - l)!} p_h^l (1 - p_h)^{m-l} = m p_h, \tag{5.65}$$

and

$$\sum_{l=0}^{m} (l - m p_h)^2 \frac{m!}{l!(m - l)!} p_h^l (1 - p_h)^{m-l} = m(1 - p_h) p_h. \tag{5.66}$$

After taking the square root, we see that typical fluctuations grow only with the square root of the number of tosses. Here we have $m = 30$ and we repeat this game $n_G = 100$ times. We bin the n_G results in 31 bins going from 0 to 30 which represent the outcomes of one game of 30 tosses. We expect that the lth bin with the number of times we get l heads in one game has a count of about $n_G p(l, m)$ occurrences with fluctuations of the order of $\sigma_l(p_h) = \sqrt{n_G p(l, m, p_h)(1 - p(l, m, p_h))}$, as the $p(l, m)$ are small, this is basically the square root of the expected number in the bin. The results of the numerical experiment in cointossing.ipynb are displayed in figure 5.9.

Are these results compatible with $p_h = 0.5$ (a fair coin)? A first test that can be performed is to count the number of heads in the 100×30 tosses and divide by 3000. In our numerical experiment it turns out to be 0.508. If the coin is fair, we expect fluctuations of the order of $\sqrt{3000 \times 0.5 \times (1 - 0.5)} \simeq 27.4$ in the total number of heads. After dividing by 3000, this turns out to be close to 0.01. Averages like 0.49 or 0.51 are fairly compatible with 0.5. A better measure of the closeness to a particular value of p_h is obtained by calculating the χ^2 of the histogram for a given p_h:

$$\chi^2(p_h) = \sum_{l=0}^{m} \left(\frac{n_l - p(l, m, p_h)}{\sigma_l(p_h)} \right)^2.$$ (5.67)

The results are shown in figure 5.10. The minimum is near $p_h = 509$ and is consistent with the average estimate.

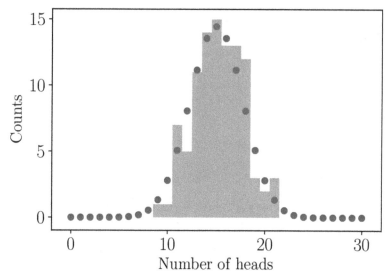

Figure 5.9. Histogram for 100 games of 30 tosses and expected values.

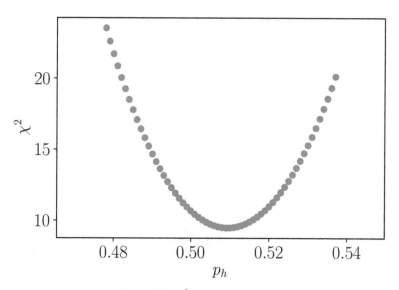

Figure 5.10. χ^2 as a function of p_h.

It is interesting to repeat the exercise with another RNG. A simple algorithm is given by the (deterministic) rule

$$x_{n+1} = ax_n \ Modulo \ m, \tag{5.68}$$

for judiciously chosen values of a and m. A very popular choice due to Lewis, Goodman, and Miller, is $a = 7^5 = 16\,807$ and $m = 2^{31} - 1 = 214\,748\,364\,7$. Another variation is to use the NIST generator of random bits from a physical process (beam splitter). See https://beacon.nist.gov/home.

cointossing

July 25, 2020

```
[1]: #import usual libraries
     import numpy as np
     import matplotlib as mpl
     import matplotlib.pyplot as plt
     %matplotlib inline
```

```
[6]: # ngames of ntoss
     ntoss=30
     ngames=100

     nheads=np.zeros(ngames)
     for gn in range(ngames):
         #we use the numpy generator
         nheads[gn]=sum(np.random.randint(0,2,ntoss))
         pass

     #using python histogram, watch for bins ending on integers!
     #https://numpy.org/doc/stable/reference/generated/numpy.histogram.html
     #calculated distribution for p=0.5
     estprob05=np.zeros(ntoss+1)
     ppp=0.5
     for nh in range(ntoss+1):
         newt=np.math.factorial(ntoss)/(np.math.factorial(nh)*np.math.
     ↪factorial(ntoss-nh))
         estprob05[nh]=(ppp**nh)*((1-ppp)**(ntoss-nh))*newt
         pass
     #comparison with p=0.5 results
     plt.hist(nheads,ntoss+1,(-0.5,ntoss+0.5),alpha=0.5)
     plt.scatter(np.linspace(0,ntoss,ntoss+1,dtype=int),ngames*estprob05)
     plt.show()
```

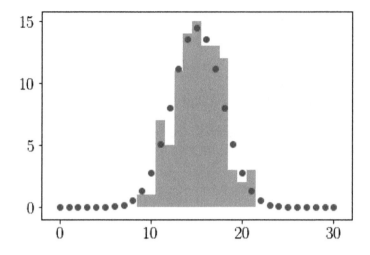

```
[7]: pyhisto=np.histogram(nheads,ntoss+1,(-0.5,ntoss+0.5))
     #total average number of heads minus the expected 0.5
     avh=sum(nheads)/(ngames*ntoss)
     print("p_h estimate:", avh)
     avsub=avh-0.5
     #estimated error for p=0.5
     exer=np.sqrt(0.25/(ntoss*ngames))
     print("expected deviation from 0.5:",exer)
     #print(avh,avsub,exer,(avh-0.5)/exer)
     #varying p and minimizing normalized chi square
     ninc=60
     #approximate best p
     guessp=np.floor(1000*avh)/1000.0
     chivspp=np.zeros((ninc,2))
     for delp in range(ninc):
         pppp=guessp-0.03+0.001*delp
         estprob=np.zeros(ntoss+1)
         ppp=pppp
         for nh in range(ntoss+1):
             newt=np.math.factorial(ntoss)/(np.math.factorial(nh)*np.math.
     ↪factorial(ntoss-nh))
             #print((ppp**nh)*((1-ppp)**(ntoss-nh))*newt)
             estprob[nh]=(ppp**nh)*((1-ppp)**(ntoss-nh))*newt
             pass
         #print(ppp,sum((myhisto[0]-ngames*estprob)**2)/30)
```

```
    chivspp[delp][0]=ppp
    chivspp[delp][1]=sum((pyhisto[0]-ngames*estprob)**2/
 ↪(ngames*estprob*(1-estprob)))
    pass
plt.scatter(chivspp[:,0],chivspp[:,1])
plt.show()
minchi=50
minp=0.0
for delp in range(ninc):
    if chivspp[delp,1]<minchi:
        minchi=chivspp[delp,1]
        minp=chivspp[delp,0]
    pass
print("minimum chi-square p_h:",minp)
```

```
p_h estimate: 0.5083333333333333
expected deviation from 0.5: 0.009128709291752768
```

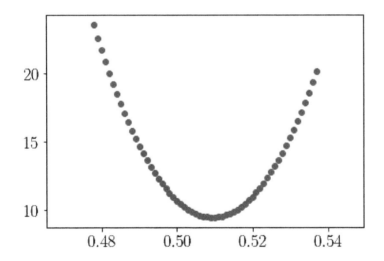

minimum chi-square p_h: 0.509

```
[8]: #plotting
     SMALL_SIZE = 14
     MEDIUM_SIZE = 16
     BIGGER_SIZE = 18

     plt.rc('font', size=MEDIUM_SIZE)            # controls default text sizes
     plt.rc('axes', titlesize=BIGGER_SIZE)       # fontsize of the axes title
     plt.rc('axes', labelsize=MEDIUM_SIZE)       # fontsize of the x and y labels
     plt.rc('xtick', labelsize=BIGGER_SIZE)      # fontsize of the tick labels
     plt.rc('ytick', labelsize=BIGGER_SIZE)      # fontsize of the tick labels
     plt.rc('legend', fontsize=SMALL_SIZE)       # legend fontsize
     plt.rc('figure', titlesize=BIGGER_SIZE)     # fontsize of the figure title
     plt.rc('axes', linewidth=1.)
     fig = plt.figure()

     plt.rc('text', usetex=True)
     plt.rc('font', family='Times')
     plt.ylabel('Counts')
     plt.xlabel('Number of heads')

     plt.hist(nheads,ntoss+1,(-0.5,ntoss+0.5),alpha=0.7)
     plt.scatter(np.linspace(0,ntoss,ntoss+1,dtype=int),ngames*estprob05,color="b")
     plt.savefig("coinstats.pdf",bbox_inches='tight')
     plt.show()
```

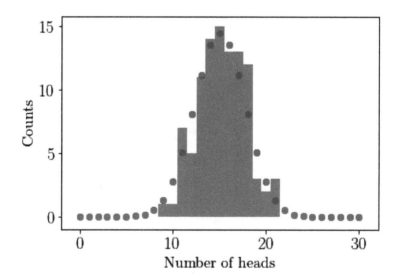

```
[9]: SMALL_SIZE = 14
     MEDIUM_SIZE = 16
     BIGGER_SIZE = 18

     plt.rc('font', size=MEDIUM_SIZE)            # controls default text sizes
     plt.rc('axes', titlesize=BIGGER_SIZE)       # fontsize of the axes title
     plt.rc('axes', labelsize=MEDIUM_SIZE)       # fontsize of the x and y labels
     plt.rc('xtick', labelsize=BIGGER_SIZE)      # fontsize of the tick labels
     plt.rc('ytick', labelsize=BIGGER_SIZE)      # fontsize of the tick labels
     plt.rc('legend', fontsize=SMALL_SIZE)       # legend fontsize
     plt.rc('figure', titlesize=BIGGER_SIZE)     # fontsize of the figure title
     plt.rc('axes', linewidth=1.)

     fig = plt.figure()

     plt.rc('text', usetex=True)
     plt.rc('font', family='Times')
     plt.ylabel('$\chi^2$')
     plt.xlabel('$p_h$')
     plt.scatter(chivspp[:,0],chivspp[:,1])
     plt.savefig("chisquare.pdf",bbox_inches='tight')
     plt.show()
```

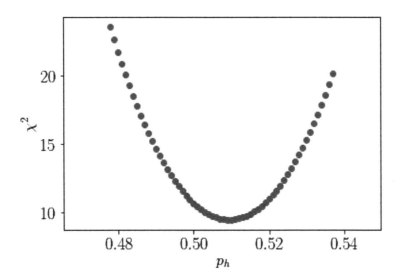

[ ]:

Exercise 11: Using importance sampling for the one-dimensional Ising model on a periodic lattice with 30 sites, calculate the average energy for $\beta = 0.5$ using 10 000 configurations. Check the results with the exact result which can be obtained by diagonalizing the transfer matrix (see chapter 9). Discuss the estimated statistical error on the sampling using a calculation of the energy fluctuations. Provide a histogram for the sample energies.

Solution. The sampling algorithm is implemented in the attached Mathematica notebook ising1.nb. The exact partition function is

$$Z = 2^L(\cosh^L \beta + \sinh^L \beta). \tag{5.69}$$

The average total energy is

$$\langle E \rangle = -\frac{d}{d\beta} \ln Z. \tag{5.70}$$

The total energy variance is

$$\sigma_E^2 = \langle (E - \langle E \rangle)^2 \rangle = \frac{d^2}{d\beta^2} \ln Z. \tag{5.71}$$

Numerically $\langle E \rangle_{\text{estimated}} = -13.74$, $\langle E \rangle_{\text{exact}} = -13.863\,5$, $\sigma_E = 4.86$. We have 10 000 configurations. The study of the correlations in the sample

$$C(t) = \sum_n (E_n - \langle E \rangle)(E_{n+t} - \langle E \rangle)/N_{\text{configs.}}, \tag{5.72}$$

indicates that the energies are decorrelated after four of the updates performed and so the estimated error is $4.86\sqrt{1/2500} \simeq 0.1$, which is compatible with the observed discrepancy.

Mathematica notebook ising.nb

```
In[124]:=  (*clear global variables*)
           Clear["Global`*"]

In[163]:=  (*fix size, temperature and boundary conditions*)
           L = 30; beta = 0.5;
           sig[0] := sig[L];
           (*energy *)
           ham := -Sum[sig[ll] * sig[ll + 1], {ll, 0, L - 1}];
           (*random spin*)
           rpm := Module[{x}, x = Random[Integer]; Return[(-1)^x]];


           (*metropolis update*)
           upgrade1 := Module[

               (*private*){del, ran, ii},

               (*here a sweep the lattice
                at random and upgrade immediately, sig is public *)

               Do[ii = Random[Integer, {1, L}]; del = 2 * sig[ii] * (sig[ii + 1] + sig[ii - 1]);
                 (*change of energy with sig[ii] flipped*)

                 If[del > 0, ran = Random[];
                   If[ran < Exp[-beta * del], sig[ii] = -sig[ii]], sig[ii] = -sig[ii]], {qq, 1, L}];

               (*final configuration after one sweep*)

               Return[Table[sig[ll], {ll, 1, L}]]];

           (*first configuration (random)*)
           Do[sig[ll] = rpm, {ll, 1, L}]

           (*can be used to see temp. conf. at any time *)

           presconf := Table[sig[ll], {ll, 1, L}];

           (* the main loop to calculate the av. energy*)
           numberofsweeps = 10 000;

           nn = 0; totalen = 0;
           While[nn ≤ numberofsweeps,
            nn = nn + 1;
            upgrade1;
            ene[nn] = ham; (*to make an histogram*)
```

```
    totalen = totalen + ham;
    If[Mod[nn, 1000] == 0, Print[N[totalen / nn, 5]]]]]
    (*to monitor*)
```

- 13.396

- 13.700

- 13.752

- 13.801

- 13.709

- 13.731

- 13.681

- 13.682

- 13.741

- 13.736

In[188]:=

```
    (* checks*)
    (*energy average from sampling*)
    aven = N[totalen / numberofsweeps]
```

Out[188]= - 13.7382

In[189]:= `(*energy average (exact)*)`
```
    averen = (-D[Log[2.^L * (Cosh[be] ^L + Sinh[be] ^L)], be]) /. {be → beta}
```

Out[189]= - 13.8635

In[190]:= `(*energy fluctuations for one sampling*)`
```
    enfluc = ((D[Log[2.^L * (Cosh[be] ^L + Sinh[be] ^L)], {be, 2}]) /. {be → beta}) ^0.5
```
Out[190]= 4.85731

In[188]:= `(*estimated error (absolute value*)`
```
    enfluc / Sqrt[numberofsweeps]
```
Out[193]= 0.0485731

In[193]= `(*actual error*)`
```
    averen - aven
```
Out[193]= - 0.125315

In[194]:= `(*energy correlations*)`

```
ListPlot[Table[
  Sum[(ene[jj] - aven) * (ene[jj + tt] - aven), {jj, 1000, 9000}] / 8000, {tt, 1, 10}]]
```

In[196]:= `(*revised error estimate*)`

```
enfluc / Sqrt[numberofsweeps / 4]
```

Out[196]= 0.0971461

In[197]:= `(*energy histogram*)`

```
histo[list_, nubin_] := Module[{newlist, mi, ma, inbin, dim, tab, xx, yy},
  Do[inbin[ll] = 0, {ll, 1, nubin + 1}];
  newlist = list;
  epsi = (Max[newlist] - Min[newlist]) / 10 000;
  mi = Min[newlist] - epsi; ma = Max[newlist] + epsi;
  dim = Dimensions[newlist][[1]];
  Do[ww = Floor[((newlist[[ll]] - mi) / (ma - mi)) * nubin] + 1;
   inbin[ww] ++
   , {ll, 1, dim}];

  (* I put the boundary of the last bin in the last bin*)
  inbin[nubin] = inbin[nubin] + inbin[nubin + 1];
  Do[xx[2 * ll] = mi + (ll) * (ma - mi) / nubin;
   xx[2 * ll + 1] = mi + (ll + 1) * (ma - mi) / nubin;
   yy[2 * ll] = inbin[ll + 1];
   yy[2 * ll + 1] = inbin[ll + 1], {ll, 0, nubin - 1}];
  xx[-1] = mi;
  yy[-1] = 0;
  xx[2 * nubin] = ma;
  yy[2 * nubin] = 0;
  tab = Table[{xx[ll], yy[ll]}, {ll, -1, 2 * nubin}];
  Return[ListPlot[tab, Joined → True, PlotRange → All]]]
```

Out[135]= 23.5934

```
In[ ]:= taben = Table[ene[ll], {ll, 1, numberofsweeps}];

In[ ]:= histo[taben, 100]
```

Exercise 12: Using importance sampling for the two-dimensional Ising model with a 32×32 lattice with periodic boundary conditions, calculate the average energy for $\beta = 0.1, 0.2, \ldots 1$. Check the results with the exact result [6].

Solution. The two methods are implemented in the attached notebooks MCIsing32. ipynb and Kaufman32.nb. The numerical results are summarized in table 5.1.

Following [6], the partition function $Z(L, \beta)$ of a finite $L \times L$ square lattice with L even and periodic boundary conditions is

$$Z(L, \beta)) = \frac{1}{2}(2 \sinh(2\beta))^{\frac{L^2}{2}} \sum_{i=1}^{4} Z_i(\beta), \qquad (5.73)$$

with

$$Z_1 = \prod_{r=0}^{L-1} 2 \cosh\left(\frac{1}{2}L\gamma_{2r+1}\right), \qquad (5.74)$$

Table 5.1. Average energies for various β with sampling (MC) and exact.

β	MC	Exact
0.1	−0.202 77	−0.203 377 4
0.2	−0.428 40	−0.428 228 8
0.3	−0.705 56	−0.704 499 1
0.4	−1.105 35	−1.107 292
0.5	−1.744 69	−1.745 565
0.6	−1.908 57	−1.909 086
0.7	−1.963 69	−1.963 776
0.8	−1.984 94	−1.984 851
0.9	−1.993 54	−1.993 494
1.0	−1.997 07	−1.997 160

$$Z_2 = \prod_{r=0}^{L-1} 2\sinh\left(\frac{1}{2}L\gamma_{2r+1}\right), \tag{5.75}$$

$$Z_3 = \prod_{r=0}^{L-1} 2\cosh\left(\frac{1}{2}L\gamma_{2r}\right), \tag{5.76}$$

$$Z_4 = \prod_{r=0}^{L-1} 2\sinh\left(\frac{1}{2}L\gamma_{2r}\right), \tag{5.77}$$

where

$$\cosh\gamma_l = c_l = \cosh(2\beta)\coth(2\beta) - \cos\left(\frac{\pi l}{L}\right), \tag{5.78}$$

so that

$$\gamma_0 = 2\beta + \log(\tanh(\beta)), \tag{5.79}$$

$$\gamma_l = \log\left(\sqrt{c_l^2 - 1} + c_l\right), \, l \neq 0. \tag{5.80}$$

These expressions contain square roots and logarithms that can lead to cuts and discontinuities if not handled properly. We know that for a finite volume, the partition function is an analytical function in the entire complex β plane, a sum of exponentials with integer weights that count the number of ways we can have a given energy. For instance, for $L = 4$, the partition function reads:

$$Z = 2e^{-32\beta} + 32e^{-24\beta} + 64e^{-20\beta} + 424e^{-16\beta} + 1728e^{-12\beta} + 6688e^{-8\beta} + 13\,568e^{-4\beta}$$
$$+ 13\,568e^{4\beta} + 6688e^{8\beta} + 1728e^{12\beta} + 424e^{16\beta} + 64e^{20\beta} + 32e^{24\beta} + 2e^{32\beta} + 20\,524.$$

As we assumed L even, the use of square roots and logs can be circumvented by expressing the factors in equation (5.77) in terms of the Chebychev polynomials:

$$\cosh((L/2)\gamma_r) = T_{L/2}(\cosh(\gamma_r)), \tag{5.81}$$

$$\sinh((L/2)\gamma_r) = U_{L/2-1}(\cosh(\gamma_r))\sinh(\gamma_r). \tag{5.82}$$

Using equation (5.78) and the pre factor to cancel poles at $\beta = 0$, we see that the first and third terms of the partition function are now polynomials of entire functions. The factors $\sinh(\gamma_r)$ have a sign ambiguity, however, they come in pairs except for $\sinh(\gamma_0)$ and $\sinh(\gamma_L)$. A careful reading of footnotes in reference [6] yields

$$\sinh(\gamma_0) = \cosh(2\beta) - \coth(2\beta), \tag{5.83}$$

$$\sinh(\gamma_L) = \cosh(2\beta) + \coth(2\beta). \tag{5.84}$$

Combining these results, the partition function is clearly an entire function. See [7] for details.

MCIsing32

July 25, 2020

```
[1]: %matplotlib inline
     import numpy as np
     import matplotlib as mpl
     import matplotlib.pyplot as plt
     import scipy.misc
     import imageio
     # provide a temperature (2.5 in example) and number of sweeps (3 in example)
     #and the strenght of the magnetic field (sth=1)
     totalsweeps=10000

     #######################
     Nx=32
     Ny=32
     nsites=Nx*Ny
     v=np.random.randint(0,2,(Nx,Ny))
     #v=np.zeros((Nx,Ny))+1
     sig=2*v-1

     for iter in range(10):
         beta=0.1+iter*0.1
         totend=0
         for nsweeps in range(totalsweeps):
             for npick in range(2*nsites):
                 nx=np.random.randint(Nx)
                 ny=np.random.randint(Ny)

     ⌴sumsig=(sig[(nx+1)%Nx,ny]+sig[(nx-1)%Nx,ny]+sig[nx,(ny+1)%Ny]+sig[nx,(ny-1)%Ny])
                 local=sig[nx,ny]*sumsig
                 if local<=0:
                     sig[nx,ny]*=(-1)
                 else:
                     pp=np.random.uniform(0,1)
                     probflip=np.exp(-2*local*beta)
                     if pp<=probflip:
                         sig[nx,ny]*=(-1)
                 pass
             #dsq=((sig+1)/2)
```

```
#hsimage=plt.imshow(dsq,cmap='Greys',aspect=1,interpolation='none')
#plt.axis('off')
#plt.show(hsimage)

men=0
for nx in range(Nx):
    for ny in range(Ny):
        men+=sig[nx,ny]*(sig[(nx+1)%Nx,ny]+sig[nx,(ny+1)%Ny])
        pass
    pass
totend+=(-men)
#if nsweeps%100==0:
 #    print(-men/(1.0*nsites))

    pass
print(beta,totend/(1.0*nsites*totalsweeps))
pass
```

```
0.1 -0.2027671875
0.2 -0.42839765625
0.30000000000000004 -0.70556015625
0.4 -1.10534765625
0.5 -1.74469453125
0.6 -1.908570703125
0.7000000000000001 -1.963693359375
0.8 -1.984941796875
0.9 -1.993541015625
1.0 -1.997071875
```

[ ]:

In[227]:=
```
Clear["Global`*"];
Print[Date[]];
n = 32; Clear[Kb]; Clear[Z]; c[r_] := Cosh[2 Kb] Coth[2 Kb] - Cos[r * π / n];
(*new formulation without square root sign ambiguity*)
ZY[1] = Product[2 * ChebyshevT[n / 2, c[2 * r + 1]], {r, 0, n - 1}];
ZY[2] = (Product[(c[2 * r + 1]) ^ 2 - 1, {r, 0, n / 2 - 1}] *
    Product[2 * ChebyshevU[n / 2 - 1, c[2 * r + 1]], {r, 0, n - 1}]);
ZY[3] = Product[2 * ChebyshevT[n / 2, c[2 * r]], {r, 0, n - 1}];
ZY[4] = (((-1 / Tanh[2 * Kb] + Cosh[2 * Kb]) * (1 / Tanh[2 * Kb] + Cosh[2 * Kb])) *
    Product[((c[2 * r]) ^ 2 - 1), {r, 1, n / 2 - 1}] *
    Product[2 * ChebyshevU[n / 2 - 1, c[2 * r]], {r, 0, n - 1}]);
ZZY = 1 / 2 (2 Sinh[2 Kb]) ^ (1 / 2 n * n) Sum[ZY[i], {i, 1, 4}];
ff = -Log[ZZY];
ee = D[ff, Kb];
dee = D[ee, Kb];
cv = - (Kb^2) * D[ee, Kb] / (n^2);
Print[Date[]];
```

{2020, 7, 25, 17, 27, 1.302499}

{2020, 7, 25, 17, 27, 2.015906}

```
(*stmp=OpenWrite["Cvexact32.d"];
Do[WriteString[stmp,datab[[ll,1]]];WriteString[stmp," "];
 WriteString[stmp,datab[[ll,2]]];
 WriteString[stmp,"\n"],{ll,1,350}];Close[stmp];*)
```

In[258]:=
```
pr = 16;
Do[
  be = SetPrecision[1/10+jj*(1/10), pr]; beta[jj] = be;
  eede[jj] = (ee /. {Kb → be}) /n^2;
  Print[{N[beta[jj], 2], N[eede[jj], 7]}], {jj, 0, 9}]
```

$$\{0.10, -0.2033774\}$$

$$\{0.20, -0.4282288\}$$

$$\{0.30, -0.7044991\}$$

$$\{0.40, -1.107292\}$$

$$\{0.50, -1.745565\}$$

$$\{0.60, -1.909086\}$$

$$\{0.70, -1.963776\}$$

$$\{0.80, -1.984851\}$$

$$\{0.90, -1.993494\}$$

$$\{1.0, -1.997160\}$$

In[260]:= `TeXForm[Table[{N[beta[jj], 2], N[eede[jj], 7]}, {jj, 0, 9}]]`

Out[260]//TeXForm=
```
\left(
\begin{array}{cc}
 0.10 & -0.2033774 \\
 0.20 & -0.4282288 \\
 0.30 & -0.7044991 \\
 0.40 & -1.107292 \\
 0.50 & -1.745565 \\
 0.60 & -1.909086 \\
 0.70 & -1.963776 \\
 0.80 & -1.984851 \\
 0.90 & -1.993494 \\
 1.0 & -1.997160 \\
\end{array}
\right)
```

5.11 Classical versus quantum

It is very common to call models formulated with the path-integral 'classical' while the corresponding formulation using a Hamiltonian acting on a Hilbert space is called 'quantum'. However, except for possible discretization artifacts, the two formulations describe the same quantum behavior. In the path-integral formulation for bosonic fields, the action is calculated in terms of c-numbers, as in the classical formulation, but the sum over all the configurations provides a quantum description which cannot be reduced to the classical solutions of the equations of motion. In other words, the path-integral is an alternate method of quantization.

In this chapter, we have started with the Hamiltonian and derived a path-integral representation. As we will see, the path-integral formalism allows formulations that are manifestly gauge-invariant and treat space and time on equal footing. For these reasons, it can be argued that the fundamental definition of a model should be done in terms of the action and the measure of integration over all the possible configurations, without reference to a Hamiltonian at the first place. The Hamiltonian can be constructed by using the transfer matrix formalism. This will be discussed in chapter 9.

A more complete discussion with detailed references of the connection between statistical mechanics in D dimensions and quantum field theoretic Hamiltonians in $D - 1$ dimensions, can be found in the classic work of Wilson and Kogut [8].

References

[1] Feynman R, Hibbs A and Styer D 2010 *Quantum Mechanics and Path Integrals* (Dover Books on Physics) (New York: Dover)

[2] Möller K 1988 *Optics* (Mill Valley, CA: University Science Books)

[3] Reed M and Simon B 1981 *I: Functional Analysis, Methods of Modern Mathematical Physics* (Amsterdam: Elsevier)

[4] Villani C 2012 *Théorème vivant, Littérature Française* (Paris: Grasset)

[5] Binder I, Heermann K, Binder K and Heermann D 2002 *Monte Carlo Simulation in Statistical Physics: An Introduction Physics and Astronomy Online Library* (Berlin: Springer)

[6] Kaufman B 1949 *Phys. Rev.* **76** 1232–43

[7] Denbleyker A, Liu Y, Meurice Y, Qin M P, Xiang T, Xie Z Y, Yu J F and Zou H 2014 *Phys. Rev.* D **89** 016008

[8] Wilson K G and Kogut J B 1974 *Phys. Rep.* **12** 75–199

IOP Publishing

Quantum Field Theory
A quantum computation approach
Yannick Meurice

Chapter 6

Lattice quantization of spin and gauge models

6.1 Lattice models

In this chapter, we will introduce lattice versions of some of the classical field theory models discussed in chapter 2. We will start with a Euclidean time and treat space and time on the same footing. The metric is simply a Kronecker delta in D dimensions. We then discretize space and time. Before discussing specific models, we introduce some of the terminology associated with the lattice.

We use a D-dimensional (hyper) cubic Euclidean space–time lattice. For instance, for $D = 2$, we use a square lattice. The *sites* are denoted $x = (x_1, x_2, \ldots x_D)$, with $x_D = \tau = it$, the Euclidean time direction. In lattice units, the space–time sites are labelled with integers

$$\frac{x}{a} = n_1 \hat{1} + \cdots + n_D \hat{D}, \tag{6.1}$$

where $n_1 \ldots, n_D$ are integers and $\hat{1}, \ldots, \hat{D}$ units vectors in the D positive directions of the (hyper) cubic lattice and a the lattice spacing. In the following, the lattice units are implicit. The basic lattice elements are shown in figure 6.1.

The *links* between two nearest neighbor lattice sites x and $x + \hat{\mu}$ are labelled by (x, μ) and the *plaquettes* delimited by four sites x, $x + \hat{\mu}$, $x + \hat{\mu} + \hat{\nu}$ and $x + \hat{\nu}$ are labelled by (x, μ, ν). By convention, we start with the lowest index when introducing a circulation at the boundary of the plaquette, as shown in figure 6.2. In enumerations of unoriented plaquettes, the condition $\mu < \nu$ is assumed, so each plaquette only appears once.

The total number of sites is denoted V. Unless specified otherwise, periodic boundary conditions are assumed, because they preserve a discrete translational symmetry. If we take the time continuum limit, we obtain a quantum Hamiltonian formulation in $D - 1$ spatial dimensions.

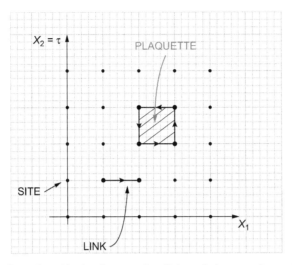

Figure 6.1. Illustration of lattice elements, sites, links and plaquettes, in two dimensions.

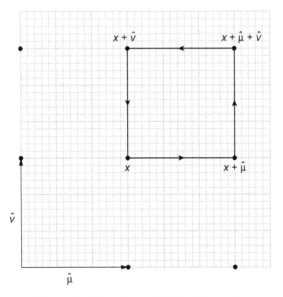

Figure 6.2. Plaquette associated with (x, μ, ν).

6.2 Spin models

At the end of section 2.4 in equation (2.47), we introduced the Lagrangian density for N real scalar fields with a $O(N)$ global symmetry. The Euclidean version reads

$$\mathcal{L}_{\text{Euclidean}}^{O(N)} = \frac{1}{2}\partial_\mu\vec{\phi} \cdot \partial_\nu\vec{\phi}\delta^{\mu\nu} + \lambda(\vec{\phi} \cdot \vec{\phi} - v^2)^2, \tag{6.2}$$

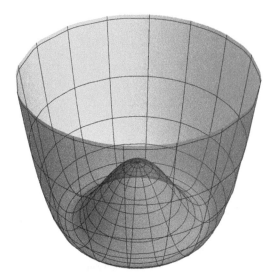

Figure 6.3. Shape of the $O(2)$-invariant potential.

with $\vec{\phi}$ a N-dimensional vector. The potential has degenerate minima on a $N-1$-dimensional hyper-sphere S_{N-1} and a local maximum at $\vec{\phi} = 0$.

For $N = 2$, the low energy part of the potential has a shape reminiscent of the bottom of a wine bottle and is displayed in figure 6.3. The degenerate minima form a circle at the very bottom. We can study the small fluctuations about a given minimum on the circle. Note that the choice of a minimum breaks the $O(2)$ symmetry. There are 'soft' fluctuations along the circle that somehow restore the symmetry and 'hard' fluctuations in the radial direction.

We can extend this analysis for arbitrary N. We pick a specific minimum of the potential pointing in the first direction and N small field fluctuations ϵ_j:

$$\phi_j = v\delta_{1j} + \epsilon_j. \tag{6.3}$$

This choice of minimum breaks the original $O(N)$ symmetry down to $O(N-1)$ (the subgroup of $O(N)$ leaving the first direction unaffected and mixing the remaining $N-1$ fields). Expending the potential up to quadratic order:

$$\lambda(\phi_j\phi_j - v^2)^2 = \lambda(2v\epsilon_1 + \epsilon_j\epsilon_j)^2 \simeq \lambda(4v^2\epsilon_1^2 + \mathcal{O}(\epsilon^3)), \tag{6.4}$$

we see that we have one massive mode (fluctuations in the first direction) with mass $2\sqrt{2\lambda}v$ and $N-1$ massless modes. The massless modes are called 'Nambu–Goldstone' (NG) modes. Their number is the number of broken generators: from the results in appendix A, $O(N)$ has $N(N-1)/2$ generators while the $O(N-1)$ residual symmetry has $(N-1)(N-2)/2$ generators. The number of broken generators is thus the difference which is $N-1$. The remaining massive mode is called the 'Brout–Englert–Higgs' (BEH) mode. If we take λ large, the BEH mode decouples from the low-energy regime and we have an effective Lagrangian for the NG modes. It is convenient to re-express the N real fields using N-dimensional hyper-spherical coordinates. The BEH mode is the radial mode while the NG modes are the $N - 1$ coordinates on S_{N-1}. In the large λ limit, we obtain a Lagrangian, which is just the kinetic term to be supplemented with the constraint that $\vec{\phi} \cdot \vec{\phi} = v^2$ everywhere.

We can write the Euclidean action for the NG modes on a D-dimensional lattice with isotropic lattice spacing a as

$$S_E = \frac{1}{2} \sum_{x,\mu} a^{D-2} (\vec{\phi}_{x+\hat{\mu}} - \vec{\phi}_x) \cdot (\vec{\phi}_{x+\hat{\mu}} - \vec{\phi}_x). \tag{6.5}$$

The constraint $\vec{\phi}_x \cdot \vec{\phi}_x = v^2$ can be expressed by introducing unit vectors: $\vec{\phi}_x = v\vec{\sigma}_x$ such that

$$\vec{\sigma}_x \cdot \vec{\sigma}_x = 1. \tag{6.6}$$

Redefining $a^{D-2}v^2 \equiv \beta$, we get the simple action

$$S = \beta \sum_{x,\mu} (1 - \vec{\sigma}_{x+\hat{\mu}} \cdot \vec{\sigma}_x). \tag{6.7}$$

These models are often called spin models or nonlinear sigma models. The first term in the action $\beta \sum_{x,\mu} 1$ is a constant that is often dropped. However, for large β, the configurations with almost constant $\vec{\sigma}_x$ dominate the partition function and since under these circumstances $\vec{\sigma}_{x+\hat{\mu}} \cdot \vec{\sigma}_x \simeq 1$, it is useful to subtract the constant in order to just keep the small fluctuations.

The case $N = 1$ is the well-known Ising model with $\sigma_x = \pm 1$. For $N = 2$, the terminology 'planar model' or 'classical XY model' is common and if we use the circle parametrization

$$\sigma_x^1 = \cos(\varphi_x), \text{ and } \sigma_x^2 = \sin(\varphi_x), \tag{6.8}$$

then

$$\vec{\sigma}_{x+\hat{\mu}} \cdot \vec{\sigma}_x = \cos(\varphi_{x+\hat{\mu}} - \varphi_x). \tag{6.9}$$

For $N = 3$, the symmetry becomes non-abelian and the model is sometimes called the 'classical Heisenberg model'. In the large-N limit, the model becomes solvable if we take the limit in such a way that $N/\beta(N) = \lambda$ remains constant [1].

6.3 Complex generalizations and local gauge invariance

It is instructive to rewrite the $O(2)$ model using the complex form

$$\Phi_x = e^{i\varphi_x}. \tag{6.10}$$

Dropping the constant term, the $O(2)$ action reads

$$S = -\frac{\beta}{2} \sum_{x,\mu} (\Phi_x^\star \Phi_{x+\hat{\mu}} + h.c.) = -\beta \sum_{x,\mu} \cos(\varphi_{x+\hat{\mu}} - \varphi_x). \tag{6.11}$$

The $O(2)$ model has a global symmetry

$$\varphi_x \rightarrow \varphi_x + \alpha. \tag{6.12}$$

With the complex notation, this transformation becomes

$$\Phi_x \rightarrow e^{i\alpha} \Phi_x. \tag{6.13}$$

We would like to promote this symmetry to a local one

$$\Phi_x \rightarrow e^{i\alpha_x} \Phi_x. \tag{6.14}$$

This can be achieved by inserting a phase $U_{x,\mu}$ between $\Phi_x^\star$ and $\Phi_{x+\hat{\mu}}$ which transforms like

$$U_{x,\mu} \rightarrow e^{i\alpha_x} U_{x,\mu} e^{-i\alpha_{x+\hat{\mu}}}. \tag{6.15}$$

The procedure can be extended for arbitrary N-dimensional complex vectors $\boldsymbol{\Phi}_x$ with a local transformation involving a $U(N)$ matrix V_x:

$$\boldsymbol{\Phi}_x \rightarrow V_x \boldsymbol{\Phi}_x. \tag{6.16}$$

In addition, we introduce $U(N)$ matrices $\mathbf{U}_{x,\hat{\mu}}$ transforming like

$$\mathbf{U}_{x,\hat{\mu}} \rightarrow V_x \mathbf{U}_{x,\hat{\mu}} V_{x+\hat{\mu}}^\dagger. \tag{6.17}$$

The action

$$S = -\frac{\beta}{2} \sum_{x,\mu} \left(\boldsymbol{\Phi}_x^\dagger \mathbf{U}_{x,\mu} \boldsymbol{\Phi}_{x+\hat{\mu}} + h.c. \right), \tag{6.18}$$

has a local $U(N)$ invariance which we call gauge invariance. In section 6.4, we show that by taking the product of a set of $\mathbf{U}_{x,\mu}$ attached to the links of closed loops, we can construct gauge-invariant quantities.

Another generalization of the $N = 1$ expression of the complex phase given in equation (6.10), consists in replacing Φ_x by a $SU(N)$ matrix $\mathbf{U}_x$. This is called the principal chiral model

$$S = -\frac{\beta}{2} \sum_{x,\mu} \left[\mathrm{tr}\left[\mathbf{U}_{x+\hat{\mu}}^\dagger \mathbf{U}_x \right] + h.c. \right]. \tag{6.19}$$

6.4 Pure gauge theories

In section 6.3, we introduced $N \times N$ unitary matrices on the links that we denoted $\mathbf{U}_{x,\mu}$. Under a gauge transformation (6.17), this matrix is multiplied on the left by V_x

and on the right by $V_{x+\hat{\mu}}^\dagger$. If we consider two successive links in positive directions, then the local transformation in the middle site cancels and

$$\mathbf{U}_{x,\mu}\mathbf{U}_{x+\hat{\mu},\nu} \rightarrow V_x\mathbf{U}_{x,\mu}\mathbf{U}_{x+\hat{\mu},\nu}V_{x+\hat{\mu}+\hat{\nu}}^\dagger. \tag{6.20}$$

If the second link goes in the negative direction, we use the Hermitian conjugate and a similar property holds

$$\mathbf{U}_{x,\mu}\mathbf{U}_{x+\hat{\mu}-\hat{\nu},\nu}^\dagger \rightarrow V_x\mathbf{U}_{x,\mu}\mathbf{U}_{x+\hat{\mu}-\hat{\nu},\nu}^\dagger V_{x+\hat{\mu}-\hat{\nu}}^\dagger. \tag{6.21}$$

We can pursue this process for an arbitrary path connecting x to some x_{final}. The transformation on the right will be $V_{x_{\text{final}}}^\dagger$. If we close the path and take the trace, we obtain a gauge-invariant quantity. We call these traces of products of gauge matrices over closed loops 'Wilson loops'. When the loop goes around the time direction, we call it a 'Polyakov loop'.

On a square, cubic or more generally hypercubic lattice, the smallest path that gives a non-trivial Wilson loop is a square. We call this square a plaquette following Claude Itzykson's suggestion at Ken Wilson's seminar in Orsay in 1973. As explained in section 6.1, a plaquette is specified by (x, μ, ν) with $\mu < \nu$. The corresponding matrix is

$$\mathbf{U}_{\text{plaquette}} = \mathbf{U}_{x,\mu\nu} = \mathbf{U}_{x,\mu}\mathbf{U}_{x+\hat{\mu},\nu}\mathbf{U}_{x+\hat{\nu},\mu}^\dagger\mathbf{U}_{x,\nu}^\dagger. \tag{6.22}$$

The simplest gauge-invariant lattice model is called Wilson's action:

$$S_{\text{Wilson}} = \beta \sum_{x,\mu<\nu} \left(1 - \frac{1}{2N}(\text{Tr}\mathbf{U}_{x,\mu\nu} + h.\ c)\right). \tag{6.23}$$

6.5 Abelian gauge models

We can write actions corresponding to the gauge part (6.23) and the gauge-matter part (6.18) for abelian groups. In the Ising case, we have

$$U_{x,\mu} = \sigma_{x,\mu} = \pm 1, \tag{6.24}$$

at every link and the pure gauge action reads

$$S_{\text{Ising gauge}} = \beta_{pl.} \sum_{x,\mu<\nu} (1 - \sigma_{x,\mu}\sigma_{x+\hat{\mu},\nu}\sigma_{x+\hat{\nu},\mu}\sigma_{x,\nu}). \tag{6.25}$$

On the gauge-matter side, we have

$$S_{\text{Ising matter}} = -\beta_{l.} \sum_{x,\mu}(\sigma_x\sigma_{x,\hat{\mu}}\sigma_{x+\hat{\mu}}). \tag{6.26}$$

In the $U(1)$ case, the unitary matrices reduce to a phase

$$U_{x,\mu} = e^{iA_{x,\mu}}, \tag{6.27}$$

and we do not need to take the trace. $A_{x,\mu}$ has a clear interpretation as a gauge field. The pure gauge action reads

$$S_{U(1)\text{ gauge}} = \beta_{pl.} \sum_{x,\mu<\nu} (1 - \cos(A_{x,\mu} + A_{x+\hat{\mu},\nu} - A_{x+\hat{\nu},\mu} - A_{x,\nu})). \qquad (6.28)$$

Under gauge transformation introduced in equation (6.17), the gauge field transformation is

$$A_{x,\mu} \rightarrow A_{x,\mu} + \alpha_x - \alpha_{x+\hat{\mu}}. \qquad (6.29)$$

For the gauge-matter part, we have

$$S_{\text{matter}} = \beta_{l.} \sum_{x,\mu} (1 - \cos(\varphi_{x+\hat{\mu}} - \varphi_x + A_{x,\mu})). \qquad (6.30)$$

This is a gauged version of the $O(2)$ model where the global symmetry under a φ shift becomes local

$$\varphi'_x = \varphi_x + \alpha_x. \qquad (6.31)$$

By combining the two actions, we obtain the compact abelian Higgs model (CAHM).

6.6 Fermions and the Schwinger model

The lattice discretization of continuous theories involving fermions involves two new features. The first one is that the Dirac field needed to be quantized by using anticommutation relations. The 'classical' version of these relations is just anticommuting fields called Grassmann variables. For N Grassmann variables ψ_i, $i = 1, 2, \ldots, N$, we have the relations

$$\{\psi_i, \psi_j\} = 0. \qquad (6.32)$$

In particular, this implies $\psi_i^2 = 0$. Grassmann numbers are used to write 'classical' actions for fields that satisfy anticommutation relations. Their integration is defined by

$$\int d\psi = 0, \text{ and } \int d\psi\psi = 1. \qquad (6.33)$$

Another new element is that the Dirac action is first order in the derivatives, as opposed to the Klein–Gordon action which is second order. This creates the fermion doubling phenomenon which can be lifted, for instance, by using a discretization due to Wilson

$$S_W = \sum_x \bar{\psi}_x \mathcal{D}_{Wxy} \psi_y, \qquad (6.34)$$

where the Wilson–Dirac operator is defined by

$$\mathfrak{D}_{Wxy} = (am + rD)\delta_{x,y} + \frac{1}{2}\sum_{\mu=1}^{D}\left\{(r - \gamma_{\mu})\delta_{y,x+\hat{\mu}} + (r + \gamma_{\mu})\delta_{x,y+\hat{\mu}}\right\}.$$

where γ_{μ} are the gamma matrices in D dimensions, and r controls the species doubling. ψ_x and $\bar{\psi}_x$ Grassmann variables with Dirac spinor indices.

There is another formulation of lattice fermions, where the different components of the fermion fields are located at different lattice sites, called staggered or Kogut–Susskind fermions [2]. The action for free fermion fields is given by,

$$S_{KS} = \frac{1}{2}\sum_{x,\mu}\eta_{x,\mu}[\bar{\psi}_x\psi_{x+\hat{\mu}} - \bar{\psi}_{x+\hat{\mu}}\psi_x], \tag{6.35}$$

where

$$\eta_{x,\mu} = (-1)^{\sum_{\nu<\mu}x_\nu} \tag{6.36}$$

with x_ν the coordinate in the νth direction. One can include gauge fields in a gauge-invariant manner by inserting $\mathbf{U}_{x,\mu}$ like,

$$S_{KS} = \frac{1}{2}\sum_{x,\mu}\eta_{x,\mu}[\bar{\psi}_x\mathbf{U}_{x,\mu}\psi_{x+\hat{\mu}} - \bar{\psi}_{x+\hat{\mu}}\mathbf{U}_{x,\mu}^{\dagger}\psi_x]. \tag{6.37}$$

A systematic discussion of lattice fermions including fermion doubling and chiral symmetry can be found in reference [3].

References

[1] David F, Kessler D A and Neuberger H 1984 *Phys. Rev. Lett.* **53** 2071–4
[2] Kogut J B and Susskind L 1975 *Phys. Rev.* D **11** 395–408
[3] Rothe H 2005 *Lattice Gauge Theories: An Introduction* (EBSCO ebook Academic Collection) (Singapore: World Scientific)

IOP Publishing

Quantum Field Theory
A quantum computation approach
Yannick Meurice

Chapter 7

Tensorial formulations

7.1 Remarks about the discreteness of tensor formulations

Before showing how the models discussed in chapter 6 can be reformulated by summing over the indices of a product of tensors, we would like to stress the mathematical reason explaining why lattice models with *compact* field variables such as the angle of the $O(2)$ model can reformulated using *discrete* indices that can be summed one by one. This is a consequence of the so-called 'Pontryagin duality' [1], which relates an abelian group and its characters (Fourier modes). It states that if the former is compact, the latter is discrete and vice versa. The simplest situation is a finite group which is compact and discrete. In the case of finite cyclic abelian groups, the characters form a finite group which is isomorphic to the group itself [2]. A simple example is $\mathbb{Z}_q$ the additive group integers modulo q. If x denotes an element of $\mathbb{Z}_q$, the characters have the form

$$\chi_k(x) = \exp(i\frac{2\pi}{q}kx), \tag{7.1}$$

and clearly satisfy the character property

$$\chi_k(x + x') = \chi_k(x)\chi_k(x'). \tag{7.2}$$

The product of two characters is another character

$$\chi_k(x)\chi_{k'}(x) = \chi_{k+k'}(x), \tag{7.3}$$

and one sees that they also form a $\mathbb{Z}_q$ group. We have studied their orthogonality relations in section 3.1.

A simple example of continuous (nondiscrete) and compact group is $U(1)$. Topologically, it is a circle and its characters are discrete and labelled by an integer (the usual Fourier modes $\exp(in\theta)$ on the circle). The orthogonality and completeness relations appear in an asymmetric way

$$\int_\pi^\pi \frac{d\varphi}{2\pi} e^{in\varphi} (e^{in'\varphi})^\star = \delta_{n,n'}, \tag{7.4}$$

and

$$\sum_{n=-\infty}^\infty e^{in\varphi} (e^{in\varphi'})^\star = 2\pi \sum_{m=-\infty}^\infty \delta(\varphi - \varphi' + m2\pi). \tag{7.5}$$

From the point of view of quantum computing, using compact fields for models with abelian symmetries guarantees a discrete set of 'states'. In the following, we will make frequent use character expansions. A simple example (for $\sigma = \pm 1$) is

$$\exp(\beta\sigma) = \cosh(\beta) + \sigma \sinh(\beta). \tag{7.6}$$

Another useful formula is the Fourier expansion

$$\exp(\beta \cos(\varphi)) = \sum_{n=-\infty}^{+\infty} I_n(\beta) \exp(in\varphi), \tag{7.7}$$

where $I_n(\beta)$ is the modified Bessel function of order n. As numerical applications require to truncate these sums, it is good to a have some idea about their convergence.

Exercise 1: Consider the approximations

$$\exp(-\beta(1 - \cos(\varphi))) \simeq \sum_{n=-n_{max}}^{+n_{max}} e^{-\beta} I_n(\beta) \exp(in\varphi), \tag{7.8}$$

that will be used in section 12.3. Plot successive approximations until they apparently converge for $\beta = 1$ and 10.

Solution. See figure 7.1.

The Peter–Weyl theorem [3] provides generalizations of Pontryagin duality to compact non-abelian groups. This will translate into expansions in spherical harmonics. The elegance and practical implications of having compact fields suggest that we should try to identify physical signatures of compactness that could allow us to test this hypothesis experimentally. This is a subject of study for the future.

Another notion of duality that we shall use is related to the Levi-Civita tensor $\epsilon^{\mu_1 \cdots \mu_D}$. Its meaning is dimension-dependent. It was already discussed in equation (2.93), where in $D = 4$, a dual field-strength tensor was defined. This duality transformation interchanges the electric and magnetic fields and reduces to the identity when repeated twice. It also relates sites to cubes and plaquettes to plaquettes. In $D = 3$, it relates the field-strength tensor to a divergenceless pseudovector $\epsilon^{ijk} \partial_j A_k$ and sites to plaquettes. This notion of duality was reviewed extensively in reference [5]. Duality is a general concept used in many branches of

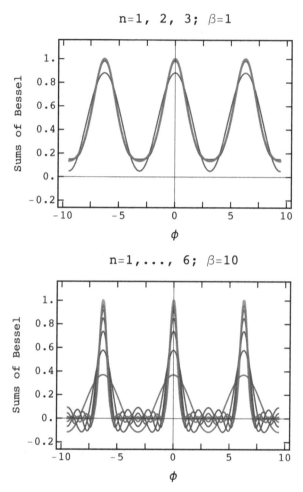

Figure 7.1. Successive approximations using equation (7.8), for $\beta = 1$ (top) and 10 (bottom). As n_{max} increases, the color changes from red to blue.

mathematics. According to Atiyah [4], duality gives 'two different points of view of looking at the same object'.

We now proceed to study spin and gauge models. We will use results from [5–7] and organize the results following the logical path of [8] sometimes with modified notations used in [9, 10].

7.2 The Ising model

The simplest application of the tensor formulation is the usual Ising model. Our first task will be to rewrite the partition function

$$Z_{\text{Ising}} = \prod_x \sum_{\sigma_x = \pm 1} e^{\beta \sum_{x,\mu} \sigma_{x+\hat{\mu}} \sigma_x}, \tag{7.9}$$

in terms of new indices attached to the links joining neighboring sites of the lattice. For each link (x, μ), we use the character expansion

$$e^{\beta \sigma_{x+\hat{\mu}} \sigma_x}$$
$$= \cosh(\beta) \sum_{n_{x,\mu}=0,1} [\sigma_{x+\hat{\mu}} \sqrt{\tanh(\beta)} \, \sigma_x \sqrt{\tanh(\beta)} \,]^{n_{x,\mu}}, \tag{7.10}$$

which attaches an index $n_{x,\mu}$ at each link (x, μ). We collect all the factors $\left(\sqrt{\tanh(\beta)} \, \sigma_x\right)^n$ from the links coming from a single site x and sum over σ_x. Using

$$\sum_{\sigma=\pm1} \sigma^n = 2\delta(\mathrm{mod}[n, 2]), \tag{7.11}$$

where $\delta(\mathrm{mod}[n, 2])$ is 1 when n is even (0 modulo 2) and 0 otherwise, we obtain an expression which depends on the indices attached at the site under consideration. This can be organized by defining a tensor. Each index can be visualized as a 'leg' coming out of the site. In D dimensions, the local tensor $T^{(x)}$ has $2D$ indices. For $D = 2$, it can be visualized as a cross. Figure 7.2 illustrates this construction. The explicit form is

$$T^{(x)}_{(n_{x-\hat{1},1}, n_{x,1}, \ldots, n_{x-\hat{D},D}, n_{x,D})}$$
$$= \sqrt{th_{n_{x-\hat{1},1}} th_{n_{x,1}}, \ldots, t_{n_{x-\hat{D},D}} th_{n_{x,D}}} \times \delta(\mathrm{mod}[n_{x,\mathrm{out}} - n_{x,\mathrm{in}}, 2]), \tag{7.12}$$

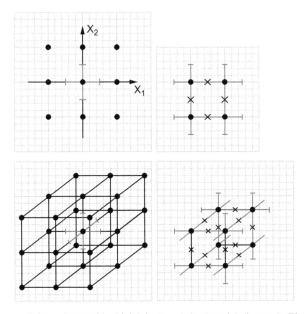

Figure 7.2. Basic tensor (left) and assembly (right) in $D = 2$ (top) and 3 (bottom). The crosses mean index contraction.

with the definitions

$$th_n \equiv (\tanh(\beta))^n$$

$$n_{x,\text{in}} \equiv \sum_\mu n_{x-\hat{\mu},\mu}$$

$$n_{x,\text{out}} \equiv \sum_\mu n_{x,\mu},$$

(7.13)

where the sums over μ run from 1 to D. We can now write the partition function as the trace of a product of tensors:

$$Z = (\cosh(\beta))^{VD} \, \text{Tr} \prod_x T^{(x)}_{(n_{x-\hat{1},1}, \, n_{x,1}, \, \ldots, \, n_{x,D})}.$$

(7.14)

The basic tensors and their assembly in two and three dimensions are illustrated in figure 7.2.

The Kronecker delta in equation (7.12) implies the discrete conservation law

$$\sum_\mu (n_{x,\mu} - n_{x-\hat{\mu},\mu}) = 0 \bmod 2,$$

(7.15)

which can be interpreted as a discrete version of a divergence-free condition. Any kind of boundary condition can be accommodated by adapting the method of integration to the link configuration at the boundary. This will be discussed in section 7.4.

7.3 $O(2)$ spin models

We now consider a generalization of the Ising model with a continuous symmetry $O(2)$. The partition function for the $O(2)$ spin model is

$$Z_{O(2)} = \prod_x \int_{-\pi}^{\pi} \frac{d\varphi_x}{2\pi} e^{\beta \sum_{x,\mu} \cos(\varphi_{x+\hat{\mu}} - \varphi_x)}.$$

(7.16)

We repeat the procedure for the Ising model now using the Fourier expansion (7.7)

$$e^{\beta \cos(\varphi_{x+\hat{\mu}} - \varphi_x)} = \sum_{n_{x,\mu}=-\infty}^{+\infty} e^{in_{x,\mu}(\varphi_{x+\hat{\mu}} - \varphi_x)} I_{n_{x,\mu}}(\beta),$$

(7.17)

where the I_n are the modified Bessel functions of the first kind. Again, this attaches an index $n_{x,\mu}$ at each link (x, μ). It is then possible to integrate over the φ_x because they appear linearly in the exponential and the integration reduces to the orthogonality relations of the Fourier modes

$$\int_{-\pi}^{\pi} \frac{d\varphi}{2\pi} e^{in\varphi} = \delta_{n,0}.$$

(7.18)

It is important to realize that the expansion (7.17) does the 'hard' work of integration exactly. The actual integration that is done after the expansion reduces to simple orthogonality relations.

We will use the translation invariant sign convention of equation (7.17) where, in the exponential, φ_x is multiplied by plus $n_{x-\hat{\mu},\mu}$ (we call them the 'in' indices) and by minus $n_{x,\mu}$ (we call them the 'out' indices). Except for the prefactor and the range of the indices, the partition function has the same form as for the Ising model:

$$Z = (I_0(\beta))^{VD} \operatorname{Tr} \prod_x T^{(x)}_{(n_{x-\hat{1},1},\, n_{x,1},\, \ldots,\, n_{x,D})}. \tag{7.19}$$

The explicit form of the local tensor $T^{(x)}$ is

$$\begin{aligned} & T^{(x)}_{(n_{x-\hat{1},1},\, n_{x,1},\, \ldots,\, n_{x-\hat{D},D},\, n_{x,D})} \\ & = \sqrt{t_{n_{x-\hat{1},1}} t_{n_{x,1}},\, \ldots,\, t_{n_{x-\hat{D},D}} t_{n_{x,D}}} \times \delta_{n_{x,\text{out}},n_{x,\text{in}}}, \end{aligned} \tag{7.20}$$

with the definitions

$$\begin{aligned} t_n &\equiv I_n(\beta)/I_0(\beta) \\ n_{x,\text{in}} &\equiv \sum_\mu n_{x-\hat{\mu},\mu} \\ n_{x,\text{out}} &\equiv \sum_\mu n_{x,\mu}, \end{aligned} \tag{7.21}$$

where the sums over μ run from 1 to D. Figure 7.2 used for the Ising model provide a graphical representations of the tensors, keeping in mind that the indices attached to the legs are now integers instead of integers modulo 2.

Exercise 2: Derive the explicit form of equation (7.20).

Solution. See a more detailed discussion of the signs in section 12.1.

Note that we factorized all the $I_0(\beta)$ factors which dominate the small β regime. We will often use the approximations

$$t_n(\beta) \equiv \frac{I_n(\beta)}{I_0(\beta)} \simeq \begin{cases} 1 - \dfrac{n^2}{2\beta} + \mathcal{O}(1/\beta^2),\ \text{for } \beta \to \infty \\[2mm] \dfrac{\beta^n}{2^n n!} + \mathcal{O}(\beta^{n+2}),\quad \text{for } \beta \to 0 \end{cases}. \tag{7.22}$$

The Kronecker delta in equation (7.20) reads

$$\sum_\mu (n_{x,\mu} - n_{x-\hat{\mu},\mu}) = 0, \tag{7.23}$$

It has the same form as equation (7.15) in the Ising case but the indices are regular integers. It is a discrete version of Noether current conservation if we interpret the $n_{x,j}$ with $j = 1, \ldots D - 1$ as spatial current densities and the $n_{x,D}$ as a charge density. At finite β, the ratios of Bessel functions t_n defined in equation (7.21) decay rapidly with n and it is justified to introduce a truncation: if any of the indices in a tensor element is larger in magnitude than a certain value n_{max}, we approximate the tensor by zero. Is this procedure compatible with the symmetries? As we will discuss in chapter 8, the answer to this question is affirmative.

The results presented above hold for the $\mathbb{Z}_q$ subgroups. The infinite sums in the character expansions are replaced by finite sums with q values. The modified Bessel functions are replaced by their discrete versions:

$$I_n(\beta) \to I_n^{(q)}(\beta) \equiv (1/q) \sum_{\ell=0}^{q-1} e^{\beta \cos(\frac{2\pi}{q}\ell)} e^{in\frac{2\pi}{q}\ell}, \tag{7.24}$$

which in the large q limit turns into the usual integral formula. We can also recover the Ising case ($q = 2$) where the general formula implies

$$I_0(\beta) \to \cosh(\beta), \text{ and } I_1(\beta) \to \sinh(\beta). \tag{7.25}$$

The selection rules in equation (7.23) remain valid modulo q.

7.4 Boundary conditions

So far we have written tensor expressions based on the assumption that each site has $2D$ neighbors. We need to discuss what needs to be done for different types of boundary conditions. Periodic boundary conditions (PBC) maintain a discrete translational invariance: the tensors themselves are the same everywhere and assembled in the same way at every site. There is no boundary and no new tensor needs to be introduced.

Open boundary conditions (OBC) can be implemented by adapting the tensor construction at the boundary. For the sites at the boundary, there are 'missing links'. The number of missing links depend on the position and the geometry. For instance for a square lattice, the sites on the sides are missing one link and the ones at the corners are missing two links. We have normalized our tensors in such a way that the tensor with $2D$ zeros $T_{00...0} = 1$. For instance, for the $O(2)$ model, we factorized $(I_0(\beta))^{VD}$ for this purpose and the cost of having an open index zero at the boundary is a weight $\sqrt{t_0} = 1$ from equation (7.20). In summary, for OBC the missing links carry an index 0 and have a weight 1. This construction can also be performed by 'decoupling' the system from a larger environment by setting β on the links connecting to this environment to zero because $t_0(0) = 1$ and $t_n(0) = 0$ when $n \neq 0$. This is illustrated in figure 7.3.

7.5 Abelian gauge theories

The character expansions used for the spin models can be adapted for gauge theories. We shall start with the simple gauge Ising model. Using the action (6.25), for each plaquette (x, μ, ν), we can write the partition function

$$Z_{\text{Ising gauge}} = \prod_{(x,\mu)} \sum_{\sigma_{x,\mu}=\pm 1} e^{\sum_{(x,\mu,\nu)} \beta \sigma_{x,\mu} \sigma_{x+\hat{\mu},\nu} \sigma_{x+\hat{\nu},\mu} \sigma_{x,\nu}}. \tag{7.26}$$

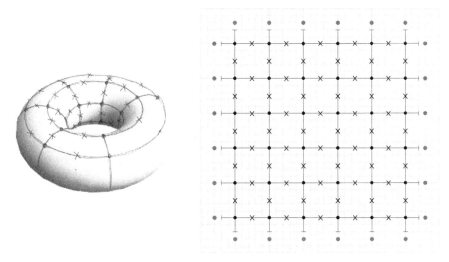

Figure 7.3. Assembling the translation invariant tensor with PBC (left), or using new tensors at the boundary for OBC (right).

As before, we expand

$$e^{\beta \sigma_{x,\mu} \sigma_{x+\hat{\mu},\nu} \sigma_{x+\hat{\nu},\mu} \sigma_{x,\nu}} = \cosh(\beta) \sum_{m_{x,\mu\nu}=0,1} [\sigma_{x,\mu} \sigma_{x+\hat{\mu},\nu} \sigma_{x+\hat{\nu},\mu} \sigma_{x,\nu} \tanh(\beta)]^{m_{x,\mu\nu}}. \tag{7.27}$$

Regrouping the factors with a given $\sigma_{x,\mu}$ and summing over ± 1, we obtain a tensor attached to this link with $2(D-1)$ indices and which enforces a selection rule

$$A^{(x,\mu)}_{m_1 \ldots m_{2(D-1)}} = \delta(\mathrm{mod}[m_1 + \cdots + m_{2(D-1)}, 2]).$$

The four links attached to a given plaquette (x, μ, ν) must carry the same index 0 or 1 attached to this plaquette in the character expansion. For this purpose, we introduce a new tensor

$$B^{(x,\mu,\nu)}_{m_1 m_2 m_3 m_4} = \begin{cases} \tanh(\beta)^{m_1}, & \text{if all } m_i \text{ are the same} \\ 0, & \text{otherwise.} \end{cases} \tag{7.28}$$

The partition function can now be written by assembling these tensors as illustrated in figure 7.4 for $D = 2$ and in figure 7.5 for $D = 3$

$$Z_{\text{Ising gauge}} = (2 \cosh \beta)^{VD(D-1)/2} \, Tr \prod_l A^{(l)}_{m_1, \ldots + m_{2(D-1)}} \prod_p B^{(p)}_{m_1 m_2 m_3 m_4}, \tag{7.29}$$

PBC provide translation invariant tensors and assembly. For OBC, we have 'missing plaquettes' for the links at the boundary. By setting the indices of the missing plaquettes at the boundary to zero, we reproduce the result obtained by direct integration.

As in the case of spin models, there is no major difficulty to extend the Ising construction to $U(1)$ or the $\mathbb{Z}_q$ subgroups. Starting with the partition function

$$Z_{U(1)} = \prod_{x,\mu} \int_{-\pi}^{\pi} \frac{dA_{x,\mu}}{2\pi} e^{\beta_{pl} \sum_{x,\mu<\nu} \cos(A_{x,\mu} + A_{x+\hat{\mu},\nu} - A_{x+\hat{\nu},\mu} - A_{x,\nu})}. \tag{7.30}$$

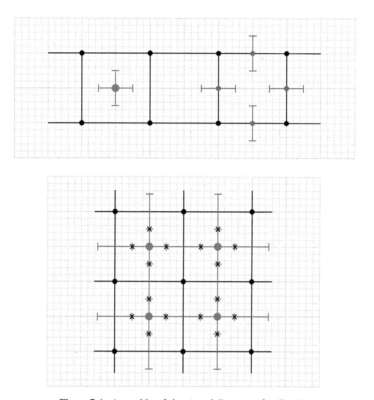

Figure 7.4. Assembly of the A and B tensors for $D = 2$.

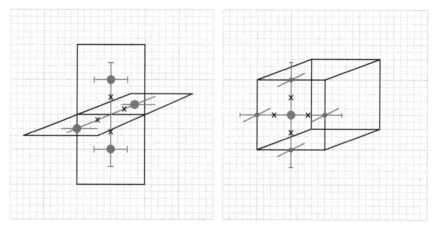

Figure 7.5. Assembly of the A and B tensors for $D = 3$.

Using the Fourier expansion

$$e^{\beta_{pl.}\cos(A_{x,\mu}+A_{x+\hat{\mu},\nu}-A_{x+\hat{\nu},\mu}-A_{x,\nu})} =$$

$$\sum_{m_{x,\mu\nu}=-\infty}^{+\infty} e^{im_{x,\mu\nu}(A_{x,\mu}+A_{x+\hat{\mu},\nu}-A_{x+\hat{\nu},\mu}-A_{x,\nu})}I_{m_{x,\mu\nu}}(\beta_{pl.}), \tag{7.31}$$

and integrating over $A_{x,\mu}$, we obtain the selection rule

$$\sum_{\nu>\mu}[m_{x,\mu\nu} - m_{x-\hat{\nu},\mu\nu}]$$

$$+ \sum_{\nu<\mu}[-m_{x,\nu\mu} + m_{x-\hat{\nu},\nu\mu}] \tag{7.32}$$

$$= 0.$$

This is is a discrete version of $\partial_\mu F^{\mu\nu} = 0$.

Exercise 3: Derive the explicit form of equation (7.32).

Solution. See a more detailed discussion of the signs in section 12.1.

As for Ising, we introduce a 'B-tensor' for each plaquette

$$B_{m_1m_2m_3m_4}^{(x,\mu\nu)} = \begin{cases} t_{m_1}(\beta_{pl.}), & \text{if all } m_i \text{ are the same} \\ 0, & \text{otherwise.} \end{cases} \tag{7.33}$$

Again they are assembled (traced) together with 'A-tensors' attached to links with $2(D-1)$ legs orthogonal to the link (x, μ)

$$A_{m_1...m_{2(D-1)}}^{(x,\mu)} = \delta_{m_{in},m_{out}}, \tag{7.34}$$

where $\delta_{m_{in},m_{out}}$ is a short notation for equation (7.32).
    The partition function with PBC reads

$$Z = (I_0(\beta_{pl.}))^{VD(D-1)/2}$$

$$\times Tr \prod_{l.} A_{m_1,...m_{2(D-1)}}^{(l.)} \prod_{pl.} B_{m_1m_2m_3m_4}^{(pl.)}, \tag{7.35}$$

where the trace means index contraction following the geometric procedure described above. As in the Ising case, tensor assembly is illustrated in figure 7.4 for $D = 2$ and in figure 7.5 for $D = 3$.

7.6 The compact abelian Higgs model

The compact abelian Higgs model (CAHM) can be constructed as a gauged version of the $O(2)$ model. In order to promote the φ shift from a global to a local symmetry,

$$\varphi_x' = \varphi_x + \alpha_x, \tag{7.36}$$

we introduced the gauge fields of the $U(1)$ model together with their plaquette interactions. The resulting partition function is

$$Z_{\text{CAHM}} = \prod_x \int_{-\pi}^{\pi} \frac{d\varphi_x}{2\pi} \prod_{x,\mu} \int_{-\pi}^{\pi} \frac{dA_{x,\mu}}{2\pi} \qquad (7.37)$$

$$e^{\beta_{pl} \sum_{pl.} \cos(A_{x,\mu} + A_{x+\hat{\mu},\nu} - A_{x+\hat{\nu},\mu} - A_{x,\nu}) + \beta_{l.} \sum_{l.} \cos(\varphi_{x+\hat{\mu}} - \varphi_x + A_{x,\mu})} \qquad (7.38)$$

As explained in section 6.2, the nonlinear $O(2)$ model is obtained by decoupling the non-compact BEH mode and only the compact NG mode is present.

The integration over φ_x leads to equation (7.23) just as for the $O(2)$ model. The integration over $A_{x,\mu}$ yields the selection rule

$$\sum_{\nu<\mu} [m_{x,\nu\mu} - m_{x-\hat{\nu},\nu\mu}] + \sum_{\nu>\mu} [-m_{x,\mu\nu} + m_{x-\hat{\nu},\mu\nu}] = n_{x,\mu}, \qquad (7.39)$$

which adds the sources $n_{x,\mu}$ in equation (7.32) and corresponds to a discrete version of Maxwell equations with charges and currents

$$\partial_\mu F^{\mu\nu} = J^\nu. \qquad (7.40)$$

Equation (7.32) implies that the link indices $n_{x,\mu}$ can be seen as completely determined by unrestricted plaquette indices $m_{x,\mu\nu}$. We write the dependence of equation (7.39) as $n_{x,\mu}(\{m\})$. Note that for $n_{x,\mu}(\{m\})$, the discrete current conservation equation (7.23) holds automatically. It is also an independent consequence of the matter field integration, however [9], there is no need to enforce equation (7.23) independently. This is a discrete version of the fact that Maxwell equations with charges and currents (7.39) imply $\partial_\mu J^\mu = 0$.

Exercise 4: Show that equation (7.39) implies equation (7.23).

We need to update the definition of the A-tensors used in the pure gauge case. The quantum numbers on the links $n_{x,\mu}$ bring a weight $t_{n_{x,\mu}}(\beta_l)$. This translates into

$$A^{(x,\mu)}_{m_1 \ldots m_{2(D-1)}} = t_{n_{x,\mu}}(\beta_l) \delta_{n_{x,\mu}, n_{x,\mu}(\{m\})}. \qquad (7.41)$$

The partition function with PBC can now be written as

$$Z_{\text{CAHM}} = (I_0(\beta_{pl.}))^{VD(D-1)/2} (I_0(\beta_{l.}))^{VD}$$
$$\times \text{Tr} \prod_{l.} A^{(l.)}_{m_1, \ldots m_{2(D-1)}} \prod_{pl.} B^{(pl.)}_{m_1 m_2 m_3 m_4}. \qquad (7.42)$$

7.7 Models with non-abelian symmetries

It is conceptually straightforward but technically tedious to extend the expansion in Fourier modes to expansions using special functions corresponding to compact

Lie groups. We discuss briefly a few examples. More details can be found in reference [8].

For the $O(3)$ nonlinear sigma model

$$Z = \prod_x \int \frac{d\Omega_x}{4\pi} \prod_{x,\mu} e^{\beta \cos \gamma_{x,\mu}}, \tag{7.43}$$

where $\cos \gamma_{x,\mu} = \vec{\sigma}_x \cdot \vec{\sigma}_{x+\hat{\mu}}$, is the angle between neighbor spins. Using the Rayleigh's formula and the addition theorem for spherical harmonics,

$$P_l(\cos \gamma_{ij}) = \frac{4\pi}{2l+1} \sum_{m=-l}^{l} Y_{lm}^*(\theta_j, \phi_j) Y_{lm}(\theta_i, \phi_i), \tag{7.44}$$

we obtain

$$e^{\beta \cos \gamma_{ij}} = \sum_{l=0}^{\infty} A_l(\beta) \sum_{m=-l}^{l} Y_{lm}^*(\theta_j, \phi_j) Y_{lm}(\theta_i, \phi_i), \tag{7.45}$$

with

$$A_l(\beta) = \frac{(2\pi)^{3/2}}{\sqrt{\beta}} I_{l+1/2}(\beta). \tag{7.46}$$

For each site, we have product of spherical harmonics. There are $2D$ of them and we expand recursively the product of two of them as sums of spherical harmonics weighted with Gaunt coefficients

$$Y_{l_1 m_1}(\theta, \phi) Y_{l_3 m_3}(\theta, \phi) = \sum_{L=l_{\min}}^{l_{\max}} G_L^{(m_1, m_3, l_1, l_3)} Y_L^{m_1+m_3}(\theta, \phi). \tag{7.47}$$

The summation bounds are

$$\begin{aligned} l_{\max} &= l_1 + l_3 \\ l_{\min} &= \begin{cases} \lambda_{\min} & \text{if } l_{\max} + \lambda_{\min} \text{ is even} \\ \lambda_{\min} + 1 & \text{if } l_{\max} + \lambda_{\min} \text{ is odd} \end{cases} \\ \lambda_{\min} &= \max(|l_1 - l_3|, |m_1 + m_3|). \end{aligned} \tag{7.48}$$

This process can be repeated until we have two sums left. The angular integration can then be performed using the orthonormal property of the spherical harmonics. For $D = 2$, the result is [8]

$$\begin{aligned} T_{(l_1, m_1), (l_2, m_2), (l_3, m_3), (l_4, m_4)} &= \delta_{m_1+m_3, m_2+m_4} \\ &\sum_L G_L^{(m_1, m_3, l_1, l_3)} G_L^{*(m_2, m_4, l_2, l_4)} \sqrt{A_{l_1} A_{l_2} A_{l_3} A_{l_4}}. \end{aligned} \tag{7.49}$$

There are now two indices associated with each leg of the tensor and the factorization of the initial tensor is lost.

Similarly for the $SU(N)$ principal chiral model we can use

$$e^{\frac{\beta}{2}Tr[U_x U_{x+\hat{\mu}}^\dagger]} = \sum_{r_{x,\mu}=0}^{\infty} F_{r_{x,\mu}}(\beta)\chi^{r_{x,\mu}}(U_x U_{x+\hat{\mu}}^\dagger), \tag{7.50}$$

and for the $SU(N)$ gauge theory

$$e^{\frac{\beta}{2}\Re Tr[U_{x,\mu\nu}]} = \sum_{r_{x,\mu\nu}=0}^{\infty} F_{r_{x,\mu\nu}}(\beta)\chi^{r_{x,\mu\nu}}(U_{x,\mu\nu}). \tag{7.51}$$

The $\chi^r(U)$ are traces for the representations indexed by r and the functions $F_r(\beta)$ are listed in reference [11] for various groups. For $SU(2)$, the matrices are Wigner functions and their product can be re-expressed as sums using Clebsh–Gordan decompositions. See reference [8] for explicit forms

7.8 Fermions

We should also mention tensorial constructions for fermions [12–14] and in particular for two-dimensional Wilson–Dirac fermions [15–17], staggered fermions for the massless Schwinger model with and without the presence of a topological term [18] and for the free Wess–Zumino model [17].

References

[1] Pontryagin L 1939 *Topological Groups* (Princeton, NJ: Princeton University Press)
[2] Serre J 1973 *A Course in Arithmetic, Graduate Texts in Mathematics* (Berlin: Springer)
[3] Peter F and Weyl H 1927 *Math. Ann.* **97** 737–55
[4] Atiyah M 2007 *Duality in Mathematics and Physics* https://fme.upc.edu/ca/arxius/butlleti-digital/riemann/071218_conferencia_atiyah-d_article.pdf
[5] Savit R 1980 *Rev. Mod. Phys.* **52** 453–87
[6] Gu Z C and Wen X G 2009 *Phys. Rev.* B **80** 155131
[7] Xie Z Y, Chen J, Qin M P, Zhu J W, Yang L P and Xiang T 2012 *Phys. Rev.* B **86** 045139
[8] Liu Y, Meurice Y, Qin M P, Unmuth-Yockey J, Xiang T, Xie Z Y, Yu J F and Zou H 2013 *Phys. Rev.* D **88** 056005
[9] Meurice Y 2019 *Phys. Rev.* D **100** 014506
[10] Meurice Y 2020 *Phys. Rev.* D **102** 014506
[11] Drouffe J M and Zuber J B 1983 *Phys. Rep.* **102** 1
[12] Gu Z C, Verstraete F and Wen X G 2010 (arXiv:1004.2563)
[13] Gu Z C 2013 *Phys. Rev.* B **88** 115139
[14] Takeda S and Yoshimura Y 2015 *Prog. Theor. Exp. Phys.* **2015** 043B01
[15] Shimizu Y and Kuramashi Y 2014 *Phys. Rev.* D **90** 014508
[16] Yoshimura Y, Kuramashi Y, Nakamura Y, Takeda S and Sakai R 2018 *Phys. Rev.* D **97** 054511
[17] Kadoh D, Kuramashi Y, Nakamura Y, Sakai R, Takeda S and Yoshimura Y 2018 *J. High Energy Phys.* **03** 141
[18] Butt N, Catterall S, Meurice Y and Unmuth-Yockey J 2019 *Phys. Rev.* D **101** 094509

IOP Publishing

Quantum Field Theory
A quantum computation approach
Yannick Meurice

Chapter 8

Conservation laws in tensor formulations

As numerical implementations or quantum simulations of lattice models with continuous symmetries require truncations, it is necessary to ask if truncations are compatible with the general identities derived from the symmetries of these models. A truncation procedure is a regularization, a way to replace an infinite sum by a finite one, and we need to figure out if the regularization is compatible with the symmetries of the theory or if it generates an 'anomaly'. As far as the universal behavior is concerned, if truncations preserve the symmetries, one should be able to obtain the properties associated with the universality classes (this concept will be discussed in chapter 11) by taking the continuum limit using a considerably simplified microscopic formulation where at each site, link or plaquettes, only a few values of the indices are kept. This is crucial when the computational units such as qubits or trapped atoms are in limited supply. These questions have been discussed in reference [1], which will be followed in the rest of this chapter.

8.1 Basic identity for symmetries in lattice models

For a lattice model with action $S[\Phi]$, with Φ a field configuration of fields ϕ_ℓ attached to locations ℓ which can be sites, links or plaquettes. Additional indices are kept implicit. The partition function has the generic form

$$Z = \int \mathcal{D}\Phi e^{-S[\Phi]}, \tag{8.1}$$

with $\mathcal{D}\Phi$ is the measure of integration over the fields. The average value of a function of the fields $f(\Phi)$ is

$$\langle f(\Phi) \rangle = \int \mathcal{D}\Phi f(\Phi)e^{-S[\Phi]}/Z. \tag{8.2}$$

Symmetries global or local can be expressed as field transformations

doi:10.1088/978-0-7503-2187-7ch8

$$\phi_\ell \rightarrow \phi'_\ell = \phi_\ell + \delta\phi_\ell[\Phi], \tag{8.3}$$

that preserve both the action and the integration measure:

$$\mathcal{D}\Phi' = \mathcal{D}\Phi \quad \text{and} \quad S[\Phi'] = S[\Phi]. \tag{8.4}$$

These symmetries form a group and the invariance is valid for any group element and not only for infinitesimal transformations. Changing variable from Φ to Φ' and using the symmetry properties of equation (8.2), we find that

$$\langle f(\Phi) \rangle = \langle f(\Phi + \delta\Phi) \rangle. \tag{8.5}$$

How can this simple expression of the symmetries be used to construct conserved quantities for global continuous symmetries as in Noether's theorem? In classical mechanics, if a transformation δq_i of generalized coordinates q_i leaves the action invariant, then after using the equation of motion, we obtain the conservation law:

$$\frac{d}{dt}\left(\frac{\partial L}{\partial \dot{q}_i}\delta q_i\right) = 0. \tag{8.6}$$

The equations of motion are obtained by taking arbitrary variations that do *not* affect the initial and final state and integrating by part the term with $\delta\frac{dq_i}{dt}$. The integration by part results in a boundary term which is zero precisely because there are no variations at the boundary. On the other hand, for a symmetry transformation δq_i is non-zero at the boundaries. In summary, if the total variation of the action is zero because it is a symmetry and if the equations of motion are used, the only terms surviving are the boundary terms. Consequently, the two individual surface terms are equal and this represents the conserved quantity.

In field theory, a similar procedure leads to a relativistically invariant current conservation

$$\partial_\mu J^\mu(x) = \vec{\nabla} \cdot \vec{J} + \frac{\partial\rho}{\partial t} = 0, \tag{8.7}$$

which is a continuity equation. By considering its integration between two time slices with spatial boundary conditions such that the spatial current does not flow outside the spatial region of integration, one obtains that the integral of the charge density over the spatial region at a given time is a constant of motion.

We will show that equation (8.5) is a global consequence of a discrete version of a continuity equation written in the local tensors. There is no need to use the equations of motion explicitly. We will first discuss the case of global continuous symmetries with an example.

8.2 The $O(2)$ model and models with abelian symmetries

For the $O(2)$ model discussed in chapters 6 and 7, the global symmetry

$$\varphi'_x = \varphi_x + \alpha. \tag{8.8}$$

implies that for a function f of M variables

$$\langle f(\varphi_{x_1}, \ldots, \varphi_{x_M}) \rangle = \langle f(\varphi_{x_1} + \alpha, \ldots, \varphi_{x_M} + \alpha) \rangle. \tag{8.9}$$

Since f is 2π-periodic in its M variables and can be expressed in terms of Fourier modes, this reduces to

$$\begin{aligned} &\langle \exp(i(n_1\varphi_{x_1} + \cdots n_M\varphi_{x_M}))\rangle \\ &= \exp(i(n_1 + \cdots n_M)\alpha)\langle\exp(i(n_1\varphi_{x_1} + \cdots n_M\varphi_{x_M}))\rangle. \end{aligned} \tag{8.10}$$

Consequently, if

$$\sum_{n=1}^{M} n_i \neq 0, \tag{8.11}$$

then

$$\langle \exp(i(n_1\varphi_{x_1} + \cdots + n_N\varphi_{x_M}))\rangle = 0, \tag{8.12}$$

because except for the phase, the left- and right-hand sides are identical.

Insertion of factors of the form $e^{in_Q\varphi_x}$ are required in order to calculate the averages (8.10). This can be done by using an 'impure' tensor at each location x which only differs from the 'pure' tensor of equation (7.20) by the Kronecker symbol replacement

$$\delta_{n_{x,\text{out}},n_{x,\text{in}}} \rightarrow \delta_{n_{x,\text{out}},n_{x,\text{in}}+n_Q}. \tag{8.13}$$

We recall that in section 7.3, we used the convention that $n_{x,\text{out}}$ was the sum of the indices $n_{x,\mu}$ coming out of x in the positive directions and multiplying $-i\varphi_x$ in the exponential, while $n_{x,\text{in}}$ was the sum of the of the indices $n_{x-\hat{\mu},\mu}$ multiplying $+i\varphi_x$. This explains why n_Q is added to $n_{x,\text{in}}$. We regarded $\delta_{n_{x,\text{out}},n_{x,\text{in}}}$ as a discrete version of Noether current conservation at a given site x. At this site, $\delta_{n_{x,\text{out}},n_{x,\text{in}}}$ is a discrete divergenceless condition and we can derive a discrete version of Gauss's theorem. If we enclose a site x in a small D-dimensional cube, the sum of indices corresponding to positive directions and coming out of the cube ($n_{x,\text{out}}$) is the same as the sum of indices corresponding to negative directions ($n_{x,\text{in}}$) coming in the cube. For instance, in two dimensions the sum of the left and bottom indices equals the sum of the right and top indices. We can 'assemble' such elementary objects by tracing over indices corresponding to their interface and construct an arbitrary domain. This is illustrated with an example in figure 8.1. From our choice of in and out indices, each tracing automatically cancels an in index with an out index and consequently, at the boundary of the domain, the sum of the in indices remains the same as the sum of the out indices.

We can now give a proof of the selection rule from a microscopic point of view. The insertion of some $e^{in_Q\varphi_x}$ amounts to inserting an 'impure' tensor defined in equation (8.13). Each insertion adds a charge n_Q, which can be positive or negative, to the sum of the out indices. We can apply this bookkeeping on an existing tensor configuration and increase the size of the enclosing box until we have gathered all

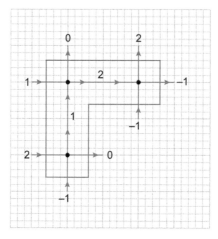

Figure 8.1. Example of flux conservation.

the insertions and reach the entire system. For PBC, we identify the boundaries and effectively remove them. Consequently, all the in and out indices get traced in pairs at the identified boundaries. This is only possible if the sum of the inserted charges is zero which is the content of equation (8.11). For OBC, all the boundary indices are zero and the same conclusion applies. In both cases, no charge can flow out of the system.

In summary, we have shown that the selection rule is a consequence of the Kronecker delta appearing in the tensor and is independent of the particular values taken by the tensors. So if we set some of the tensor elements to zero as we do in a truncation, this does not affect the selection rule.

The argument extends for discrete $\mathbb{Z}_q$ subgroups and in particular for the Ising model. The only differences are that the phases resulting from shifts involve $2\pi/q$ angles and the sums are understood modulo q. More specifically [1], the infinite sums in the Fourier expansions are replaced by finite sums with q values and the modified Bessel functions are replaced by their discrete counterparts:

$$I_n(\beta) \rightarrow I_n^{(q)}(\beta) \equiv (1/q) \sum_{\ell=0}^{q-1} e^{\beta \cos\left(\frac{2\pi}{q}\ell\right)} e^{in\frac{2\pi}{q}\ell}, \qquad (8.14)$$

which in the large q limit turns into the usual integral formula. In the Ising case $(q = 2)$, we have

$$I_0(\beta) \rightarrow \cosh(\beta), \quad \text{and} \quad I_1(\beta) \rightarrow \sinh(\beta). \qquad (8.15)$$

8.3 Non-abelian global symmetries

For the $O(3)$ model, we can expand the Boltzmann weights in terms of spherical harmonics. For a global rotation R, equation (8.10) becomes

$$\langle Y_{\ell_1 m_1}(\theta_{x_1}, \varphi_{x_1}) \dots Y_{\ell_N m_N}(\theta_{x_N}, \varphi_{x_N}) \rangle = D^{\ell_1}_{m_1 m_1'}(R)$$
$$\dots D^{\ell_N}_{m_N m_N'}(R) \langle Y_{\ell_1 m_1'}(\theta_{x_1}, \varphi_{x_1}) \dots Y_{\ell_N m_N'}(\theta_{x_N}, \varphi_{x_N}) \rangle,$$

where the $D^{\ell}_{mm'}(R)$ are the matrices corresponding to the ℓ representation and the m_i' indices are summed from $-\ell_i$ to ℓ_i. By using iteratively the Clebsch–Gordan decomposition, the expectation value can be expressed into a sum of irreducible representations. Only the singlets are allowed to get a non-zero expectation value. Arbitrary truncations are likely to generate non-zero expectation values for the non-singlets. However, if in any given irreducible representation ℓ we keep or discard *all* the m's between $-\ell$ and ℓ, then we preserve the global rotational invariance. This can be seen from the spherical harmonic expansion of the Boltzmann weights. If two neighbor spins are specified by unit vectors with spherical coordinates θ_i, φ_i and θ_j, φ_j, respectively, and if we denote the angle between them γ_{ij} we have

$$\cos(\gamma_{ij}) = \sin(\theta_i) \sin(\theta_j) \cos(\varphi_i - \varphi_j) + \cos(\theta_i) \cos(\theta_j). \tag{8.16}$$

Using Rayleigh formula and the identity

$$P_\ell(\cos(\gamma_{ij})) = \frac{4\pi}{2\ell + 1} \sum_{m=-\ell}^{\ell} Y^{\star}_{\ell m}(\theta_j, \varphi_j) Y_{\ell m}(\theta_i, \varphi_i), \tag{8.17}$$

we find that

$$e^{\beta \cos \gamma_{ij}} = \sum_{l=0}^{\infty} A_l(\beta) \sum_{m=-l}^{l} Y^{*}_{lm}(\theta_j, \varphi_j) Y_{lm}(\theta_i, \varphi_i), \tag{8.18}$$

with

$$A_l(\beta) = (2\pi)^{\frac{3}{2}} \beta^{-\frac{1}{2}} i^{l+\frac{1}{2}} J_{\ell+\frac{1}{2}}(-i\beta). \tag{8.19}$$

This shows that the truncation in ℓ respects the global symmetries because

$$\sum_{m=-\ell}^{\ell} Y^{\star}_{\ell m}(\theta, \varphi) Y_{\ell m}(\theta', \varphi'), \tag{8.20}$$

is invariant under global rotations. It seems possible to extend the argument beyond this special example.

Unlike Noether's standard field theoretical construction, our construction does not rely on taking infinitesimal symmetry transformations. The character expansions require the full group. Consequently, everything we did applies to discrete subgroups. For global non-abelian symmetries, the truncation must keep a certain number of irreducible representations and combine the weights in a way that is manifestly invariant before the field integrations are performed, as we showed explicitly for the $O(3)$ sigma model. It seems possible to extend this construction in more general circumstances.

8.4 Local abelian symmetries

We can imitate the procedure of section 7.3 and assign 'in' and 'out' qualities to the legs of the A-tensors in abelian gauge theories with respect to the link where they are attached. For a given site x and a pair of directions μ and ν, there are eight coplanar legs for the A-tensors, attached to the links in the $\pm\hat{\mu}$ and $\pm\hat{\nu}$ directions, which are themselves attached to x. We label these eight coplanar legs $[(x, \mu), \pm\hat{\nu}]$, $[(x - \hat{\mu}, \mu), \pm\hat{\nu}]$, $[(x, \nu), \pm\hat{\mu}]$, and $[(x - \hat{\nu}, \nu), \pm\hat{\mu}]$. The pair of indices appearing first refers to the link where the A-tensor is attached and the second index to the direction of the leg which can be positive or negative. The $[(x, \mu), \hat{\nu}]$ with $\mu < \nu$ are given an out assignment. There are three operations that swap in and out: changing (x, μ) into $(x - \hat{\mu}, \mu)$, changing $\hat{\mu}$ into $-\hat{\mu}$ and interchanging μ and ν. Using these three commuting operations, we can generate the eight legs. These eight legs can be assembled in a plaquette centered at the site x. This is not a plaquette of the original lattice but rather a plaquette of the dual lattice in the plane considered and where the centers of the B tensors are located. There are four B tensors at the corners of this plaquette. For $D = 2$, we get the simple picture shown in figure 8.2.

It is useful to identify the in and out assignments in this simple two-dimensional example. Our starting point is the $[(x, 1), \hat{2}]$ leg which comes out vertically from the positive horizontal axis. We now consider the three operations that transform this out-leg into three in-legs. $[(x, 1), -\hat{2}]$ comes in from below the positive horizontal axis. $[(x - \hat{1}, 1), \hat{2}]$ comes in from above the negative side of the horizontal axis. $[(x, 2), \hat{1}]$ comes in from the right of the positive vertical axis. We can then apply two different operations and obtain three out-legs: $[(x - \hat{1}), -\hat{2}]$ coming out down from the negative horizontal axis, $[(x, 2), -\hat{1}]$ coming out left on the positive vertical axis and $[(x - \hat{2}, 2), \hat{1}]$ coming out right of the negative vertical axis. Finally, if the three operations are applied, we get $[(x - \hat{2}, 2), -\hat{1}]$ coming in left of the negative vertical

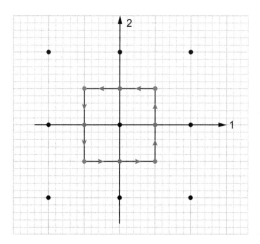

Figure 8.2. A-tensor redundancy for $D = 2$.

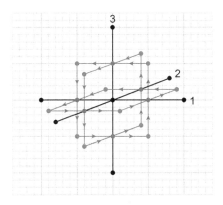

Figure 8.3. A-tensor redundancy for $D = 3$.

axis. From inspection, the eight legs form a circuit with a current circulating counterclockwise.

We see that the assignment gives consistent in–out assignments at the B tensors. In other words, at each B-tensor, there is one in-leg and one out-leg. For $D = 2$, the delta at the four links is a divergenceless condition in $D - 1$ dimension. One of these four is redundant because it follows from the three others. This has a simple local symmetry interpretation: we can use the local symmetry at x to gauge one of the links coming out of x to the identity. After the gauge fixing, there is no integration over the link and so no corresponding A-tensor. However, its effect is a consequence of the three other A-tensors. The construction will hold for any pair of directions in higher dimensions.

The reasoning can be extended to higher dimensions. The case $D = 3$ is illustrated in figure 8.3. There are now three pairs of directions and three mutually orthogonal plaquettes centered at x and their boundary can fit on the boundary of a cube. The six A-tensors are on the faces of the cube. The in-out assignments of the legs of each A-tensor are fixed and connect consistently at the B-tensor locations. We can now use the same method as in the case of the discrete Gauss's theorem. By assembling five A-tensors we cover the cube except for one face. As the B-tensors do not change the quantum numbers, we obtain the quantum numbers of the sixth A-tensor and they satisfy the divergenceless condition. Again this can be interpreted as gauge fixing the link orthogonal (piercing) the sixth A-tensor.

Exercise 1: Discuss the generalization of the above construction for $D = 4$.

8.5 Generalization of Noether's theorem

In the previous subsection, we connected a local symmetry at a site x with the redundancy of a tensor attached to a link coming out of x. We can actually extend the reasoning for global symmetries with PBC or OBC. If we consider the proof of the discrete Gauss's theorem without charge insertions, we see that charge does not flow out of space–time with **PBC** or **OBC**. Consequently, if we remove one site, the quantum numbers associated with the legs of the tensor coming out of this site are completely determined by the quantum numbers of all the other tensors. Consequently, we can remove the Kronecker delta associated with this tensor, or equivalently use the global symmetry to set the field variable at this site to zero. Following reference [2] where additional examples are discussed, we can reformulate Noether's theorem for global, local, continuous or discrete abelian symmetries: for each given symmetry, there is one corresponding tensor redundancy.

References

[1] Meurice Y 2019 *Phys. Rev.* D **100** 014506
[2] Meurice Y 2020 *Phys. Rev.* D **102** 014506

IOP Publishing

Quantum Field Theory
A quantum computation approach
Yannick Meurice

Chapter 9

Transfer matrix and Hamiltonian

The transition from the Lagrangian formulation considered in the previous chapters, to the quantum Hamiltonian formulation can be achieved by using the transfer matrix [1–4]. For a lattice model with N_τ sites in the Euclidean time direction we re-express the partition function as

$$Z = \int \mathcal{D}\Phi \, e^{-S[\Phi]_E} = \text{Tr}(\mathbb{T}^{N_\tau}). \tag{9.1}$$

The transfer matrix can be used to define a Hamiltonian by taking an anisotropic limit where β becomes large on time links or space–time plaquettes and we obtain an expression of the form

$$\mathbb{T} \propto (e^{-a_\tau H}), \tag{9.2}$$

with a_τ the lattice spacing in the time direction. As a consequence, the Hamiltonian will inherit the properties of the transfer matrix. On the practical side, the construction of the transfer matrix can be accomplished easily with the tensor formulation. The basic idea is to postpone the summation over the indices associated with links pointing in the time direction, or plaquettes with one time and one space direction. We now discuss separately the spin and gauge cases. Following the discussion at the end of section 8.2, the results for discrete subgroups of continuous groups can be obtained easily from the continuous results and we will first discuss the $O(2)$ and $U(1)$ cases.

9.1 Transfer matrix for spin models

For spin models [5], the transfer matrix is constructed by tracing over the spatial indices of all the tensors on a time slice (see figure 9.1 for $D = 2$ and $D = 3$). With PBC or OBC, there is no flow of indices 'leaking' in the spatial directions. Consequently, given the global conservation law discussed in section 8.2, the sum of the time indices going in the time slice equals the sum of the indices going out.

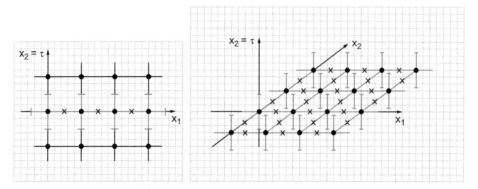

Figure 9.1. Illustration of the $O(2)$ transfer matrix in two (left) and three (right) dimensions.

This conserved quantity can be identified as the charge of the initial and final states. The transfer matrix commutes with the charge operator which counts the sum of the in or out indices. As explained in section 8.2, the symmetry is encoded in the selection rule and setting some matrix elements to zero if some of the local indices exceed some value n_{max} in absolute value will not affect this property.

Note that it is possible to introduce a chemical potential which couples to the conserved quantity by shifting the differences of angles in the time direction by $-i\mu$. Four instance for the $O(2)$ model, the Fourier expansion becomes

$$e^{\beta \cos(\varphi_{x+\hat{\tau}}-\varphi_x-i\mu)} = \sum_{n_{x,\tau}=-\infty}^{+\infty} e^{in_{x,\tau}(\varphi_{x+\hat{\mu}}-\varphi_x)}e^{\mu n_{x,\tau}}I_{n_{x,\tau}}(\beta). \tag{9.3}$$

In the same way as taking the derivative with respect to β inserts minus the energy in the partition function, taking the derivative with respect to the chemical potential inserts the charge.

We now concentrate on the $O(2)$ model. The Hilbert space $\mathfrak{H}$ is the product of integer indices attached to time links between two time slices

$$\mathfrak{H} = |\{n\}\rangle = \otimes_{\mathbf{x},j}|n_{\mathbf{x},j}\rangle. \tag{9.4}$$

For $D = 2$ with N_s sites and PBC in space, the matrix elements of the transfer matrix $\mathbb{T}$ have the explicit form

$$\langle\{n'\}|\mathbb{T}|\{n\}\rangle = \mathbb{T}_{(n_1,n_2,\dots,n_{N_S})(n_1',n_2',\dots,n_{N_S}')}$$
$$= \sum_{\tilde{n}_1\tilde{n}_2\cdots\tilde{n}_{N_S}} T^{(1,\tau)}_{\tilde{n}_{N_S}\tilde{n}_1n_1n_1'}T^{(2,\tau)}_{\tilde{n}_1\tilde{n}_2n_2n_2'}\cdots T^{(N_S,\tau)}_{\tilde{n}_{L_x-1}\tilde{n}_{N_S}n_{N_S}n_{N_S}'}. \tag{9.5}$$

The explicit form of the tensors is provided in section 7.3.

We take the time continuum limit [2, 3] by setting

$$u = 1/(\beta_\tau a_\tau), \quad \tilde{\mu} = \mu/a_\tau \text{ and } J = \beta_s/a_\tau, \tag{9.6}$$

and taking the limit $a_\tau \to 0$. This defines the Hamiltonian

$$\mathbb{H} = \frac{u}{2}\sum_{\mathbf{x}} \hat{L}_{\mathbf{x}}^2 - \tilde{\mu}\sum_{\mathbf{x}} \hat{L}_{\mathbf{x}} - \frac{J}{2}\sum_{\mathbf{x},\mu}(\hat{U}_{\mathbf{x}+\hat{\mu}}\hat{U}_{\mathbf{x}}^{\dagger} + \text{h. c. }). \tag{9.7}$$

The last term involves

$$\hat{U} \equiv \widehat{e^{i\varphi}}, \tag{9.8}$$

and is the operator version of the cosine of the difference of the angles in the spatial direction. The operator $\widehat{e^{i\varphi}}$ corresponds to the insertion of $e^{i\varphi_x}$ in the path integral and raises the charge. A similar construction appears in the Schwinger model [6]. We now discuss the commutation relations following reference [7].

For $D = 1$ the correspondence between the path integral and the operator formalism is easy to understand because there is only one way for the charge to flow. The tensor has the form

$$T_{n_x,n_{x-1}} = t_{n_x}(\beta)\delta_{n_x,n_{x-1}}, \tag{9.9}$$

and represents a diagonal transfer matrix (there is only one spatial site). For large β, $t_n(\beta) \simeq 1 - n^2/2\beta$ and we identify the time lattice spacing with $1/\beta$. The rotor spectrum has energies $E_n = n^2/2$. The value of the conserved charge n is the angular momentum of the rotor. For PBC, the partition function is the trace of the N_τ power of the transfer matrix. If we insert $e^{i\varphi_x}$ in the functional integral, the charge n increases by 1 and the trace is zero unless we insert $e^{-i\varphi_{x+y}}$ or a product of insertions having the same effect. In order to setup the Hamiltonian formalism, we introduce the angular momentum eigenstates

$$\hat{L}|n\rangle = n|n\rangle,$$
$$\hat{H}|n\rangle = \frac{n^2}{2}|n\rangle, \tag{9.10}$$

with n taking any integer value from $-\infty$ to $+\infty$. As $\hat{H} = (1/2)\hat{L}^2$,

$$[\hat{L}, \hat{H}] = 0, \tag{9.11}$$

and the angular momentum eigenstates are also energy eigenstates. The insertion of $e^{i\varphi_x}$ in the path integral becomes an operator $\widehat{e^{i\varphi}}$ which raises the charge

$$\widehat{e^{i\varphi}}|n\rangle = |n + 1\rangle, \tag{9.12}$$

while its Hermitian conjugate lowers it

$$(\widehat{e^{i\varphi}})^{\dagger}|n\rangle = |n - 1\rangle. \tag{9.13}$$

We obtain the commutation relations

$$[L, \widehat{e^{i\varphi}}] = \widehat{e^{i\varphi}}, [L, \widehat{e^{i\varphi}}^{\dagger}] = -\widehat{e^{i\varphi}}^{\dagger}, \tag{9.14}$$

and

$$[\widehat{e^{i\varphi}}, \widehat{e^{i\varphi}}^{\dagger}] = 0. \tag{9.15}$$

We now discuss the effect of a truncation on these algebraic results. By truncation we mean that there exists some n_{max} for which

$$\widehat{e^{i\varphi}}|n_{max}\rangle = 0, \text{ and } (\widehat{e^{i\varphi}})^\dagger | - n_{max}\rangle = 0. \tag{9.16}$$

If we consider the commutation relation with this restriction, we see that the only changes are

$$\langle n_{max}|[\widehat{e^{i\varphi}}, \widehat{e^{i\varphi}}^\dagger]|n_{max}\rangle = 1,$$
$$\langle -n_{max}|[\widehat{e^{i\varphi}}, \widehat{e^{i\varphi}}^\dagger]| - n_{max}\rangle = -1, \tag{9.17}$$

instead of 0. The important point is that the truncation does not affect the basic expression of the symmetry in equation (9.11) but only affects matrix elements involving the $\widehat{e^{i\varphi}}$ operators. A related discussion of the algebra for the $O(3)$ model can be found in reference [8]. Other deformations of the original Hamiltonian algebra appear in the quantum link formulation of lattice gauge theories [9] where $\widehat{e^{i\varphi}}$ is replaced by the raising operator of the $SU(2)$ algebra in a representation of dimension $2n_{max} + 1$. Equation (9.15) is then replaced by the standard $SU(2)$ commutation relation between raising and lowering operators. It should also be noticed that equations (9.14) and (9.15) correspond to the $M(2)$ algebra, the rotations and translations in a plane. Its representations are infinite dimensional with matrix elements given in terms of Bessel functions [10].

Note that we call $\widehat{e^{i\varphi}}$ and $\widehat{e^{i\varphi}}^\dagger$ raising and lowering operators, however, unlike for the standard harmonic oscillator algebra, there is no lowest state annihilated by the lowering operator. Such a state would be in contradiction with the requirement (9.15) that $\widehat{e^{i\varphi}}$ and $\widehat{e^{i\varphi}}^\dagger$ commute, because if $\widehat{e^{i\varphi}}^\dagger|lowest\rangle = 0$, we obtain the contradiction

$$0 = \widehat{e^{i\varphi}}\widehat{e^{i\varphi}}^\dagger|lowest\rangle = \widehat{e^{i\varphi}}^\dagger\widehat{e^{i\varphi}}|lowest\rangle = \widehat{e^{i\varphi}}^\dagger|lowest + 1\rangle = |lowest\rangle \neq 0.$$

A few more remarks about algebra representations in truncated spaces. So far, we have discussed situations involving compact field variables because they provide natural ways to discretize the transfer matrix. Recently [11–13], Gaussian quadratures have been used to discretize the integration of ϕ^4 theories. This procedure is closely related to the idea of realizing the harmonic oscillator algebra with a state with maximal energy 'annihilated' by the creation operator. To see this, it is useful to recall that the recursion relation for the Hermite polynomials is a consequence of the harmonic oscillator commutation relation

$$[a, a^\dagger] = 1, \tag{9.18}$$

realized using

$$a^\dagger|n\rangle = \sqrt{n+1}|n+1\rangle \text{ and } a|n\rangle = \sqrt{n}|n-1\rangle. \tag{9.19}$$

If we define

$$\langle x|n \rangle \equiv \frac{1}{\sqrt{2^n n!}} \frac{1}{\pi^{1/4}} e^{-x^2/2} H_n(x), \tag{9.20}$$

then

$$\langle x|\hat{X}|n \rangle = x\langle x|n \rangle = \frac{1}{\sqrt{2}}(\langle x|n+1 \rangle \sqrt{n+1} + \langle x|n-1 \rangle \sqrt{n}).$$

This is equivalent to the Hermite polynomial recursion relation

$$H_{n+1}(x) - 2xH_n(x) + 2nH_{n-1}(x) = 0. \tag{9.21}$$

A way to terminate the process at level n_{max} is to restrict x to values x_j such that

$$H_{n_{max}+1}(x_j) = 0, \tag{9.22}$$

with $j = 1, \ldots, n_{max} + 1$ labelling the zeros of $H_{n_{max}+1}(x)$. This is precisely the way the sampling values for the Gaussian quadrature approximation are picked. At the algebraic level we can define modified creation and annihilation operators $\tilde{a}^\dagger$ and $\tilde{a}$ which act on the finite Hilbert $|0\rangle, \ldots, |n_{max}\rangle$ as in equation (9.20) except that we have

$$\tilde{a}^\dagger|n_{max}\rangle = 0. \tag{9.23}$$

In addition, the modified commutation relation become

$$[\tilde{a}, \tilde{a}^\dagger] = \mathbb{1} - (n_{max} + 1)|n_{max}\rangle\langle n_{max}|. \tag{9.24}$$

Note that the modified operators can be represented as $(n_{max} + 1) \times (n_{max} + 1)$ matrices and the trace of their commutator is zero, as is the trace of the right-hand side.

The same method can be used to realize the truncated algebra (9.17). If y denotes the eigenvalues of $(\widehat{e^{i\varphi}} + \widehat{e^{i\varphi}}^\dagger)/2$, then we obtain the Chebyshev polynomial recursion relation. More specifically, if we define $\langle y|n \rangle = C_n(y)$, we obtain

$$C_{n+1}(y) - 2yC_n(y) + C_{n-1}(y) = 0. \tag{9.25}$$

If we start with $C_{-n_{max}} = 1$ and $C_{-n_{max}-1} = 0$, we can use the Chebyshev polynomials of the second kind and $C_n(y) = U_{n+n_{max}}(y)$ for the wavefunctions. We then impose that $C_{n_{max}+1}(y_j) = 0$ for the selected y_j which are the zeros of $U_{2n_{max}+1}$. Using the known representation

$$U_n(y) = \frac{\sin((n+1)\arccos(y))}{\sin(\arccos(y))}, \tag{9.26}$$

the selected values are

$$y_j = \cos(j\pi/(2(n_{max} + 1))), \tag{9.27}$$

for $j = 1, \ldots, 2n_{max} + 1$.

9.2 Gauge theories

For gauge theories, we can organize the tensor trace by assembling 'time layers' corresponding to 'magnetic' time slices and 'electric' slices half-way between the magnetic time slices. Clearly, this singles out a time direction as needed for the Hamiltonian treatment.

In the following we focus on the CAHM and the pure gauge limit can be obtained be decoupling the matter field. For the CAHM, the case $D = 2$ is discussed in reference [14] and generalized for arbitrary dimension in reference [15] which will be followed hereafter. The pure gauge $D = 3$ case is discussed in reference [16].

For $D = 3$, the transfer matrix can be visualized as a 'lasagne' with alternating magnetic and electric two-dimensional layers. The magnetic time slices contain B-tensors on space–space plaquettes and the A-tensors attached to their space links. In D dimensions, these A-tensors have $2(D - 2)$ legs in spatial directions, which will be assembled with other A tensors using B tensors, and two legs in opposite time directions which we can identify with the 'past' and the 'future'. See figure 9.2 in the case $D = 3$. These legs in the time direction will be connected to the B tensors in space–time plaquettes and belonging to the electric layers.

In between the magnetic time slices there are electric layers with B-tensors on space–time plaquettes labelled by $e_{(x,\tau),j}$ with a fixed Euclidean time τ, and the A-tensors attached to their time links. These A-tensors have $2(D - 1)$ legs all in spatial directions. See figure 9.3 for $D = 3$. When seen 'from above', without the time legs of the B-tensors, this looks like the $D = 2$ assembly for the $O(2)$ model.

Following reference [15], these two types of layers can be represented as matrices connecting electric states with an Hilbert space

$$|\{\mathbf{e}\}\rangle = \otimes_{\mathbf{x},j}|e_{\mathbf{x},j}\rangle. \tag{9.28}$$

We use this basis because the B-tensors on the space–time plaquettes have two legs in the time direction. In this basis, the electric layer is a diagonal matrix $\mathbb{T}_E$ with matrix elements

$$\langle\{\mathbf{e}'\}|\mathbb{T}_E|\{\mathbf{e}\}\rangle = \delta_{\{\mathbf{e}\},\{\mathbf{e}'\}}T_E(\{\mathbf{e}\}), \tag{9.29}$$

where $T_E(\{e\})$ can be written with some implicit notations as a traced product of A tensors on time links with B tensors on space–time plaquettes

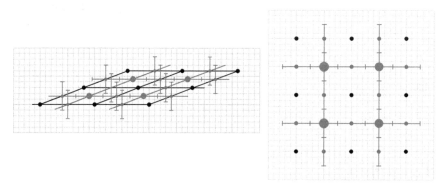

Figure 9.2. Magnetic layer of the transfer matrix for $D = 3$ in a time slice (left) and 'from above' without the time legs of the A-tensor (right).

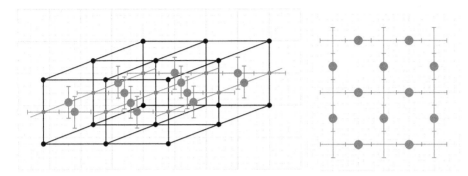

Figure 9.3. Electric layer of the transfer matrix for $D = 3$ between two time slices (left) and 'from above' (right).

$$T_E(\{\mathbf{e}\}) = \mathrm{Tr} \prod_{\text{time } l.} A^{(l.)}_{m_1, \ldots m_{2(D-1)}} \prod_{\text{sp.-time pl.}} B^{(\text{pl.})}(\mathbf{e}). \tag{9.30}$$

In a similar way, we define a magnetic matrix $\mathbb{T}_M$ such that $\langle\{e\}|\mathbb{T}_M|\{e'\}\rangle$ with the indices e and e' carried by the time legs of the A-tensors.

$$\langle\{\mathbf{e}'\}|\mathbb{T}_M|\{\mathbf{e}\}\rangle = \mathrm{Tr} \prod_{\text{sp.}l.} A^{(l.)}_{m_1, \ldots m_{2(D-1)}}(\mathbf{e}, \mathbf{e}') \prod_{\text{sp.-sp.pl.}} B^{(\text{pl.})}. \tag{9.31}$$

All the traces are over the spatial legs, while the time legs provide the matrix indices e and e'. In figures 9.2 and 9.3 the horizontal lines correspond to traced indices while the vertical indices carry the $\{e\}$ indices. The transfer matrix $\mathbb{T}$ is defined as

$$\mathbb{T} \equiv (e^{-\beta_{\text{pl.}}}I_0(\beta_{\text{pl.}}))^{(V/N_\tau)D(D-1)/2}(e^{-\beta_{l.}}I_0(\beta_{l.}))^{(V/N_\tau)D} \times \mathbb{T}_E^{1/2}\mathbb{T}_M\mathbb{T}_E^{1/2}, \tag{9.32}$$

and satisfies the basic requirement (9.1).

Note that $\mathbb{T}_E$ only involves time links and plaquettes having one direction in time. Following [2, 3], we introduce separate β_τ couplings for $\mathbb{T}_E$ and use redefinitions in terms of the time lattice spacing a_τ:

$$\beta_{\tau\text{pl.}} = \frac{1}{a_\tau g_{\text{pl.}}^2}, \text{ and } \beta_{\tau l.} = \frac{1}{a_\tau g_{l.}^2}. \tag{9.33}$$

Given the weak coupling (large β) behavior of $t_n(\beta)$ given in equation (7.22), at first order in a_τ, we get 'rotor' energies $(1/2)g_{\text{pl.}}^2 m^2$ for the plaquettes and $(1/2)g_{l.}^2 n^2$ for the links. On the other hand, $\mathbb{T}_M$ only involves space links and space–space plaquettes and we use

$$\beta_{s\,\text{pl.}} = a_\tau J_{\text{pl.}}, \text{ and } \beta_{s\,l.} = a_\tau h_{l.}. \tag{9.34}$$

Given the strong coupling (small β) behavior of $t_n(\beta)$ from equation (7.22), at first order in a_τ, the contribution to $\mathbb{T}_M$ involves a *single* link or plaquette with quantum number ± 1 all the other ones having a quantum number 0 and a weight 1. This leaves us with only few options: raise or lower $e_{x,j}$ over a link (x, j) or raise over two links and lower over the two other links of a plaquette. The Hamiltonian $\mathbb{H}$ is defined as the order a_τ correction to the identity in the transfer matrix:

$$\mathbb{T} = 1 - a_\tau \mathbb{H} + \mathcal{O}(a_\tau^2). \tag{9.35}$$

As in section 9.1, we introduce operators $\hat{e}_{\mathbf{x},j}$ and $\hat{U}_{\mathbf{x},j}$ such that

$$\begin{aligned}
\hat{e}_{\mathbf{x},j}|e_{\mathbf{x},j}\rangle &= e_{\mathbf{x},j}|e_{\mathbf{x},j}\rangle \\
\hat{U}_{\mathbf{x},j}|e_{\mathbf{x},j}\rangle &= |e_{\mathbf{x},j}+1\rangle \\
\hat{U}_{\mathbf{x},j}^\dagger|e_{\mathbf{x},j}\rangle &= |e_{\mathbf{x},j}-1\rangle.
\end{aligned} \tag{9.36}$$

The discussion of the first order behavior of $\mathbb{T}_E$ and $\mathbb{T}_M$ in the lattice spacing implies that

$$\begin{aligned}
\mathbb{H} =\ & \frac{1}{2}g_{\text{pl.}}^2 \sum_{\mathbf{x},j}(\hat{e}_{\mathbf{x},j})^2 \\
& + \frac{1}{2}g_{l.}^2 \sum_{\mathbf{x}}\left(\sum_j (\hat{e}_{\mathbf{x},j} - \hat{e}_{\mathbf{x}-\hat{j},j})\right)^2 \\
& - h_{l.} \sum_{\mathbf{x},j}(\hat{U}_{\mathbf{x},j} + h.\,c.) \\
& - J_{\text{pl.}} \sum_{\mathbf{x},j<k}(\hat{U}_{\mathbf{x},j}\hat{U}_{\mathbf{x}+\hat{j},k}\hat{U}_{\mathbf{x}+\hat{k},j}^\dagger\hat{U}_{\mathbf{x},k}^\dagger + h.\,c.).
\end{aligned} \tag{9.37}$$

The reasons we recover the Kogut–Susskind (KS) Hamiltonian [3] and how it also appears in different contexts [17, 18] are discussed in reference [15]. We have used

$$\sum_{j=1}^{D-1}(e_{x,j} - e_{x-\hat{j},j}) = n_{x,D}, \tag{9.38}$$

to eliminate $n_{x,D}$. This is a discrete version of Gauss's law and we can use it to eliminate the quantum numbers associated with the matter scalar field. In other words, we can use these quantum numbers to absorb the non-zero divergence of the electric quantum numbers. However, in the pure gauge limit, the matter fields decouple and we need to impose the condition on the Hilbert space.

9.3 $U(1)$ pure gauge theory

We can take the pure gauge limit by setting $\beta_l = 0$ in equation (7.38). From equation (7.41), we find that $t_{n_{x,\mu}}(0)$ is nonzero only if Gauss's law without charge density is satisfied. In the rest of this subsection, 'Gauss's law' is understood without charge density, a discrete version of $\nabla \cdot E = 0$. It reads

$$\sum_{j=1}^{D-1}(e_{x,j} - e_{x-\hat{j},j}) = 0. \tag{9.39}$$

This imposes a restriction on the Hilbert space which becomes equivalent to the set of legal tensor configurations of a $D - 1$ dimensional $O(2)$ model. The integer quantum numbers of the space–time plaquettes $e_{x,j}$ are like the link variables $n_{x,j}$ for the $O(2)$ model. Both sets of indices are divergenceless.

As we see from equation (9.39), Gauss's law implies that the $D - 1$ components of $e_{x,j}$ are not independent and that we could reduce the size of the Hilbert space if we could eliminate the redundant degrees of freedom. It should also be noted that if we apply $\mathbb{T}_M$ on a state that satisfies Gauss's we get another state satisfying Gauss's law. This is actually a consequence of the discrete Maxwell's equations (7.32).

Exercise 1: Prove this statement.

Solution. See section III.C of [15].

For a noisy quantum computer, the exact preservation of Gauss's law by the time evolution can be partially destroyed by a certain number of computer errors. This provides an additional reason to have Gauss's law automatically satisfied. In the rest of this subsection, we follow the approach of reference [15]. More insight can be found in references [16, 19–22].

Gauss's law can be automatically satisfied by introducing a new set of quantum numbers $c_{x,j,k}$, associated with the plaquettes of a $D - 1$ lattice. They are not related to the existing gauge quantum numbers. For an arbitrary configuration $\{c_{x,j,k}\}$, this can be accomplished in the following way:

$$
\begin{aligned}
e_{x,j} &= \sum_{k>j}[-c_{x,j,k} + c_{x-\hat{k},j,k}] \\
&\quad + \sum_{k<j}[c_{x,k,j} - c_{x-\hat{k},k,j}].
\end{aligned}
\tag{9.40}
$$

This is a discrete version of

$$
E^k = \partial_j C^{jk},
\tag{9.41}
$$

For an arbitrary antisymmetric tensor C^{jk} with space indices j, k running from 1 to $D - 1$. It is possible to introduce dimension-dependent 'magnetic' notations such as $G = \epsilon^{kl} C^{kl}$ for $D = 3$ and $G^j = \epsilon^{jkl} C^{kl}$ for $D = 4$.

For a $D = 3$ pure gauge theory we can think about the electric Hilbert space as a $D = 2$ $O(2)$ model being on a plane between two time slices, as in figure 9.3. We can further visualize the new variables as located in the middle of the plaquettes of this 'horizontal' plane, which means in the center of the $D = 3$ cubes of the original lattice. This is equivalent to the dual formulation discussed in reference [16].

For $D = 4$, the expression for the electric field is a discrete version of

$$
\mathbf{E} = \nabla \times \mathbf{G}.
\tag{9.42}
$$

This implies Gauss's law, but $\nabla \times E$ is in general non-zero which is why we don't use this method for conventional electrostatics [23].

This method is optimal for $D = 3$, because it reduces the dimensionality of the Hilbert space to one index per site ($c_{x,1,2}$) rather than 2 ($e_{x,1}$ and $e_{x,2}$). For $D = 4$, there are three indices per sites in both case, because $c_{x,j,k}$ is only defined up to a gradient. The way we can use this freedom to reduce the size of the Hilbert space depends on the boundary conditions and is discussed in reference [15, 22]. When the Hilbert space is parametrized with the new quantum numbers in equation (9.40), the relation between the $e_{x,j}$ and $c_{x,j,k}$ is linear. We can study the effect of changing one of the $c_{x,j,k}$ by ± 1. For instance, $\Delta c_{x,1,2} = 1$ generates the following changes:

$$\Delta e_{x,1} = -1, \ \Delta e_{x+\hat{2},1} = 1, \ \Delta e_{x,2} = 1, \ \Delta e_{x+\hat{1},2} = -1. \tag{9.43}$$

This change can be visualized as an electric field circulating clockwise on a plaquette in the 1–2 plane and it clearly satisfies Gauss's law. The changes correspond to the $U^{\dagger}U^{\dagger}UU$ term in the Kogut-Susskind Hamiltonian. For $D = 3$, we can efficiently replace the term with two raising and two lowering operators by a term with a single raising or lowering operator [16]. The construction can be repeated for any pair of directions in higher dimensions, but the $c_{x,j,k}$ have some redundancy. For $D = 4$, the geometric interpretation is that we can combine six plaquettes on a cube in such a way that all the electric quantum numbers cancel. In other words, the effect of one of the $c_{x,j,k}$ can also be obtained with five others. For OBC, we could remove this redundancy by eliminating, for instance, all the $c_{x,2,3}$ except for those on a 2–3 plane at the boundary. For PBC, other sectors should be added in order to allow electric configurations wrapping around the spatial directions.

Exercise 2: Consider a $U(1)$ pure gauge theory in 2 + 1 dimensions on a 3 by 3 spatial lattice (four plaquettes) with OBC. Using equation (9.40), calculate the 12 electric quantum numbers on the links as a function of the 4 $c_{x,1,2} \equiv c_x$. Check that Gauss's law is satisfied at the nine sites.

Solution. Use

$$e_{x,1} = -c_x + c_{x-\hat{2}}, \text{ and } e_{x,2} = c_x - c_{x-\hat{1}}, \tag{9.44}$$

with c_x is non-zero inside the four plaquettes and zero outside, as shown in figure 9.4.

9.4 Historical aspects of quantum and classical tensor networks

In the previous chapters, we used 'classical tensors' to reformulate 'classical lattice models' using discrete summations. In this chapter, we connected to the 'quantum Hamiltonian' and used tensor networks on spatial lattices. Historically, the opposite process took place. In the early 2000s, the idea of describing quantum states, operators and partition functions by assembling some basic tensors, which can be represented by connected 'legs' carrying indices, was pursued [24–30] in various

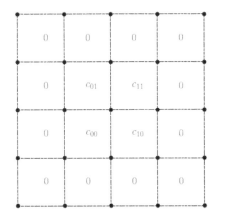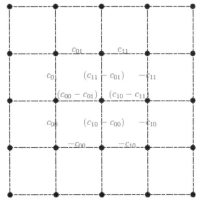

Figure 9.4. Illustration of the solution for the divergenceless electric field (red on the links) in terms of the c_x field (blue in the plaquettes).

contexts related to the density matrix renormalization group (DMRG) method [31] (for a review see [32]). One important idea appearing in the above references is the representation of quantum states by matrix product states (MPS).

As an example, we consider a one-dimensional quantum chain problem with N_s sites. An arbitrary element of the Hilbert space Hilbert space can be written as

$$|\psi\rangle = \sum_{i_1,\ldots,i_{N_s}} c_{i_1,\ldots,i_{N_s}} |i_1, \ldots, i_{N_s}\rangle, \tag{9.45}$$

where

$$|i_1, \ldots, i_{N_s}\rangle = |i_1\rangle \otimes \cdots \otimes |i_{N_s}\rangle, \tag{9.46}$$

and each of the indices runs over a local Hilbert space of dimension d_H and attached to a site. For instance, for the spin-1/2 Heisenberg model, $d_H = 2$. The dimension of the N_s-site Hilbert space is $d_H^{N_s}$, an exponential growth with the size of the system that rapidly becomes computationally unmanageable. In the MPS approach, one assumes the form

$$c_{i_1,\ldots,i_{N_s}} = \text{Tr}[A_{i_1}\ldots A_{i_{N_s}}], \tag{9.47}$$

where the A_{i_j} are $d_B \times d_B$ matrices for each value of i_j. A graphical representation of such state is provided in figure 9.5. The filled circles represent the matrices, the vertical lines represent open indices with d_H values and the horizontal lines traced indices with d_B values.

It can be shown that if for a given N_s, we take d_B large enough then we can obtain an arbitrarily good approximation of the original states. In order for this representation to be useful, we need to be able to capture the physics of large N_s with a reasonably small and N_s-independent d_B. If so, the size of the sub-Hilbert space used for computations is only growing like $N_s \times d_B^2 \times d_H$, so linearly with the size of the system. In a similar way, one can represent operators using the trace of $N_s d_B \times d_B$

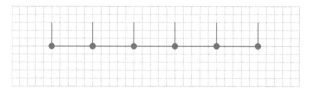

Figure 9.5. Illustration of states in the MPS approach.

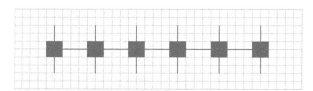

Figure 9.6. Illustration of operators in the MPS approach.

matrices $A_{i_j}^{i'_{j'}}$ with two indices in the one-site Hilbert spaces at a computational cost scaling like $N_s \times d_B^2 \times d_H^2$. An MPS operator is illustrated in figure 9.6.

At about the same time, it was realized [33, 34] that tensorial representations could also be used to describe the partition functions and expectation values for classical statistical mechanics or field theory models. This is basically the approach that we have been following here. The same authors used singular value decomposition (SVD) for 'rewiring' and sum over closed loops in order to get a coarse-grained version of the original model. They called this approach the tensor renormalization group (TRG). A more complete set of references can be found in reference [35]. The coarse-graining aspects will be discussed in chapter 11. The TRG approach was combined with higher order SVD (HOTRG) to produce accurate numerical calculations [36] for the classical Ising model. It was realized that this approach offered great advantages for lattice field theory [37] compared to the approximations of Migdal [38] and Kadanoff [39] and that character expansions [19, 40] developed in strong coupling approaches, could be used in a generic way to deal with models studied in lattice gauge theory, as shown in reference [41] and discussed here.

9.5 From transfer matrix functions to quantum circuits

The use of the transfer matrix to construct the Hamiltonian suggests a more general computational picture where tensorial methods can be used to construct circuits that can be implemented on quantum computers. The tensorial representation of the partition function of a spin model is depicted in the left side of figure 9.7. We now go back from Euclidean to real time by taking β_τ purely imaginary. If we start with an approximation where all the spatial couplings are set to zero (right side of figure 9.7) we have a set of decoupled quantum rotors and a very simple real-time evolution. As illustrated at the bottom of figure 9.7, we can then reintroduce the spatial interactions in a sequential way with controllable errors using the Trotter formula.

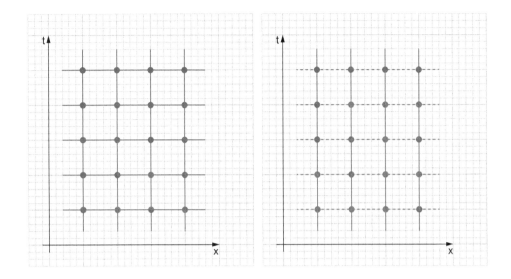

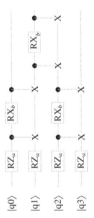

Figure 9.7. Graphical representation of the partition function for a spin model like the $O(2)$ model (left) and in the isolated rotors approximation (right). The dash lines stand for a single index 0 rather than a sum over all the states represented by straight lines. Circuit for 4 qubits with open boundary conditions used in reference [42] (bottom). Reprinted with permission of APS from [42].

Adapting the S Lloyd argument [43] for a a system composed of N quantum rotors with an energy truncation approximation (each isolated rotor is described by a finite Hilbert space of dimension d_H), we can write the Hamiltonian as

$$H = \sum_{j=1}^{\ell} H_j, \tag{9.48}$$

where each H_j acts on a space that involves either a single rotor or two neighbor rotors. It is clear that ℓ increases linearly with N but the connectivity is fixed by the dimension and independent of N. The Trotter error associated with the approximation

$$e^{-iHt} \simeq (e^{-iH_1 t/n} \dots e^{-iH_\ell t/n})^n + \cdots \tag{9.49}$$

can be controlled by taking n large enough as discussed in section 5.4. The size of the Hilbert space is d_H^N and writing and manipulating matrices takes a classical computer time growing exponentially fast with N. However, if the Hilbert space can be implemented with actual physical devices and if the evolution for the individual H_j can be performed on the Hilbert space for one or two rotors, while keeping the coherence of the wave function, the quantum computing time scales linearly in N and t. The tensor representation can be used to build a sequence of independent evolutions acting on one or two rotors. An example of a quantum circuit for the Ising model that will be discussed in the next subsection is shown figure 9.7.

9.6 Real time evolution for the quantum ising model

We now specialize the discussion of the real-time evolution to the quantum Ising model. Adapting the discussion leading to equation (9.7) to the Z_2 subgroup for a spatial lattice in $D - 1$ dimensions, we find that

$$\hat{H}_{\text{Ising}} = -h_T \sum_{\mathbf{x}} \hat{\sigma}_{\mathbf{x}}^x - J \sum_{\langle \mathbf{x}, \mathbf{y} \rangle} \hat{\sigma}_{\mathbf{x}}^z \hat{\sigma}_{\mathbf{y}}^z, \tag{9.50}$$

with $\langle x, y \rangle$ labelling pairs of nearest neighbors for any kind of boundary conditions. The Hilbert space can be represented as the tensor product of eigenstates of the Pauli matrix $\hat{\sigma}_x^z$:

$$\mathfrak{H} = \otimes_{\mathbf{x}} |{\pm}1\rangle_{\mathbf{x}}. \tag{9.51}$$

The second term of the Hamiltonian corresponds to the classical interactions in a given time slice. The connection to the classical theory is clear in this basis which we call the 'spin basis'.

We now specialize the discussion to one spatial dimension with three types of boundary conditions. The quantum Hamiltonian with N_s sites is defined as

$$\hat{H}_{\text{spin}} = -h_T \sum_{j=1}^{N_s} \hat{\sigma}_j^x . -J \left(\sum_{j=1}^{N_s-1} \hat{\sigma}_j^z \hat{\sigma}_{j+1}^z + b \hat{\sigma}_1^z \hat{\sigma}_{N_s}^z \right), \tag{9.52}$$

with $b = 0$ for open boundary conditions (OBC), $b = 1$ for periodic boundary conditions (PBC) and $b = -1$ for antiperiodic boundary conditions (ABC). The Hilbert space has dimension 2^{N_s}.

In the following we will rather use a representation where the first term, the transverse magnetic field term, is diagonal because it allows us to see the connection between field theory and quantum mechanics more easily. We call this basis the 'particle basis.' In this basis, the Hamiltonian reads

$$\hat{H}_{\text{part.}} = -h_T \sum_{j=1}^{N_s} \hat{\sigma}_j^z - J \left(\sum_{j=1}^{N_s-1} \hat{\sigma}_j^x \hat{\sigma}_{j+1}^x + b \hat{\sigma}_1^x \hat{\sigma}_{N_s}^x \right). \tag{9.53}$$

The two representations are connected by a site-wise Hadamard unitary transformation. Again, the Hilbert space is equivalent to N_s qubits. The particle basis was used in reference [42] to study the real time evolution of the model using quantum computers with up to 8 qubits.

We are now ready to carry out the time evolution in units where $\hbar = 1$ using the evolution operator

$$U(t) = e^{-it\hat{H}_{\text{part.}}}. \tag{9.54}$$

In the following, we will focus on site occupation observables for states

$$|\psi(t)\rangle = U(t)|\psi(0)\rangle, \tag{9.55}$$

with $|\psi(0)\rangle$ in what we will call the one-particle sector. The motivation for using the particle basis is that when the nearest neighbor interactions J in the spatial directions are small, there is a simple physical picture where qubit states can be interpreted as approximate particle occupations. Using exact diagonalization, for initial states with one or two particles, it was shown [42] and we will show below that for small J, discrete Bessel functions provide very accurate expressions for the evolution of the occupancies corresponding to initial states with one and two particles with specific boundary conditions. Boundary conditions play an important role when the evolution time is long enough. Using a simple laptop, it is easy to perform exact diagonalization for $N_s \sim 10$. On the other hand the discrete Bessel function can be used in good approximation for small J and much larger systems.

In the particle basis, we define the 'particle number' at each j site as

$$\hat{n}_j = (1 - \hat{\sigma}_j^z)/2. \tag{9.56}$$

We will use these quantum numbers to classify the states of the Hilbert space which is a direct product of two-dimensional (qubit) spaces at each of the N_s spatial sites. As an example for $N_s = 4$, if we label the qubits as 0, 1, 2, and 3 form left to right, the action of a sample operator on a sample state can be illustrated as

$$\hat{\sigma}_0^x |1001\rangle = |0001\rangle. \tag{9.57}$$

We now introduce the raising and lowering operators

$$\hat{\sigma}_j^{\pm} \equiv \frac{\hat{\sigma}_j^x \pm i\hat{\sigma}_j^y}{2}. \tag{9.58}$$

Note that with our occupation number (qubit) convention $|0\rangle$ corresponds to the eigenstate $+1$ of $\hat{\sigma}^z$ and consequently $\hat{\sigma}_j^-|0\rangle = |1\rangle$ and $\hat{\sigma}_j^+|1\rangle = |0\rangle$. It is easy to think that the eigenstate $+1$ of $\hat{\sigma}^z$ is the ground state of a magnetic Hamiltonian but we need to remember that if we use the standard definition of raising and lowering operators used for the angular momentum algebra, the 'raising operator' brings the excited state into the ground state.

We can write

$$\hat{\sigma}_j^x \hat{\sigma}_{j+1}^x = (\hat{\sigma}_j^+ + \hat{\sigma}_j^-)(\hat{\sigma}_{j+1}^+ + \hat{\sigma}_{j+1}^-) \tag{9.59}$$

$$= \hat{\sigma}_j^+ \hat{\sigma}_{j+1}^- + \hat{\sigma}_j^- \hat{\sigma}_{j+1}^+ + \hat{\sigma}_j^+ \hat{\sigma}_{j+1}^+ + \hat{\sigma}_j^- \hat{\sigma}_{j+1}^-. \tag{9.60}$$

The first two terms conserve the total particle number $\hat{n} = \sum_i \hat{n}_i$ and correspond to the hopping into neighbor sites. The third term destroys pairs and the fourth term creates pairs. The Hamiltonian only connects states for which the total particle number is the same modulo 2. A more systematic discussion of the symmetries of the model can be found in reference [42].

In the limit $J = 0$, the energy is the sum of the on-site energies. There is a unique ground state where all sites have an energy $-h_T$ and so $E^{(0)} = -N_s h_T$. Next, we have degenerate 'one-particle' states where one on-site state with energy $+h_T$ can be placed at N_s locations. If the h_T energy is located at the site j, we call this state $|j\rangle$. These states have an energy $-(N_s - 2)h_T$. The effect of the nearest neighbor interactions can be included perturbatively or treated exactly by performing a Wigner–Jordan transformation [3]. When $J \ll h_T$, we can in first approximation ignore the effect of pair creation in the one-particle sector and just consider the particle hopping that stays in the one-particle sector. It is worth noting that the approximate particle conservation makes this model in the small J limit similar to the XY model which has been studied thoroughly. A discussion of the literature can be found in reference [42]. If periodic boundary conditions are imposed, Fourier modes diagonalize the perturbation. This lifts the degeneracy by a term proportional to $-2J\cos(2\pi m/N_s)$. The perturbation also contains operators that connect to the three-particle states; this leads to energy shifts $\mathcal{O}(J^2/h_T)$. If we neglect these second order effects, we have a simple approximate quantum mechanical behavior. We can then prepare the system in an initial state $|\psi\rangle$ and calculate $\langle\psi(t)|\hat{n}_l|\psi(t)\rangle$, where the calculations in the quantum mechanical approximation are relatively easy. For instance, for $|\psi(0)\rangle = |j\rangle$, we obtain

$$\langle\psi_j(t)|\hat{n}_l|\psi_j(t)\rangle \simeq |J_{l-j}^{(N_s)}(2Jt)|^2, \tag{9.61}$$

where the discrete Bessel functions are defined as,

$$J_n^{(N_s)}(x) = \frac{(-i)^n}{N_s} \sum_{m=0}^{N_s-1} e^{i(\frac{2\pi mn}{N_s} + x\cos(\frac{2\pi m}{N_s}))} \tag{9.62}$$

which is the usual definition in the limit of large N_s.

Exercise 3: Derive equation (9.61).

Solution. In the one-particle approximation, we can replace $\hat{n}_l$ by $|l\rangle\langle l|$ and we obtain that

$$\langle \psi_j(t)|\hat{n}_l|\psi_j(t)\rangle \simeq |\langle l|\psi_j(t)\rangle|^2 = |\langle l|e^{-i\hat{H}t}|j\rangle|^2. \tag{9.63}$$

In the one-particle approximation, $\hat{H}$ is the sum of shift forward plus a shift backward. For PBC, the eigenstates of the shift are the momentum states

$$\langle j|m\rangle = \frac{1}{\sqrt{N_s}}e^{i\frac{2\pi}{N_s}jm}, \tag{9.64}$$

with $m = 0, 1, \ldots, N_s - 1$. The corresponding eigenvalues for a forward shift are $e^{i\frac{2\pi}{N_s}m}$ and the eigenvalues of $\hat{H}$ are $-2J\cos(\frac{2\pi}{N_s}m)$ plus a constant. Inserting the identity as $\sum_m |m\rangle\langle m|$ in equation (9.63) we get the Bessel function up to a phase which disappears when we multiply by the complex conjugate.

In the following, we propose to carry on some numerical exercise related to the real-time evolution of the quantum Ising model. We urge the readers to develop their own code and check in various ways suggested below. This code can also be used to reproduce the graphs in the two following subsections.

Exercise 4: Setting up the Hilbert space. For N_s sites (qubits), consider the integers between 0 and $2^{N_s} - 1$. Define two functions (or use the ones available in suitable libraries), one mapping the integer into its binary decomposition and the other being its inverse. These will be used to manipulate and enumerate N_s-qubit states. For instance for $N_s = 4$, $|0000\rangle \leftrightarrow |0\rangle$, $|0001\rangle \leftrightarrow |1\rangle\ldots$, $|1111\rangle \leftrightarrow |15\rangle$. Construct N_s sets of Pauli matrices $\vec{\sigma}_j$ acting exclusively on the jth qubit $j = 0, 1, \ldots N_s - 1$. Use $|\ldots0\ldots\rangle$ as $\begin{pmatrix} 1 \\ 0 \end{pmatrix}$ and $|\ldots1\ldots\rangle$ as $\begin{pmatrix} 0 \\ 1 \end{pmatrix}$. For $N_s = 4$, $\sigma_2^x|0011\rangle = |0001\rangle$ and $\sigma_3^z|0011\rangle = -|0011\rangle$ are examples. Check that the known algebraic relations are satisfied: the Pauli matrices associated with the same site anticommute while those for different sites commute.

Exercise 5: Setting up the Hamiltonian. Consider $N_s = 4$ with $H_T = -\sum_{j=1}^{N_s}\sigma_j^z$ and $H_{NN} = -\sum_{j=1}^{N_s-1}\sigma_j^x\sigma_{j+1}^x$. Calculate the eigenvectors and eigenvalues of H_T, H_{NN} and $H = H_T + 0.02H_{NN}$. H_T and H_{NN} can be diagonalized exactly. H is the quantum Ising model with $h_t = 1$ and $J = 0.02$.
Solution. For H_T the ground state is $|0000\rangle$ with four spins up, the first excited state has a degeneracy four $|1000\rangle$, $|0100\rangle$, $|0010\rangle$, $|0001\rangle$. Proceeding similarly for the rest of the states, we obtain the spectrum: $-4, -2, -2, -2, -2, 0, 0, 0, 0, 0, 0, 2, 2, 2, 2, 4$. For H_{NN}, we use the basis where σ^x is diagonal and obtain $-3, -3, -1, -1, -1, -1, -1, -1, 1, 1, 1,$ $1, 1, 1, 3, 3$ from the enumeration of domains (each wall costs an amount of energy of 2. For H, we obtain numerically $-4.000\,300\,00$, $-2.032\,577\,04$, $-2.012\,442\,89$, $-1.987\,722\,96$, $-1.967\,857\,12$, $-4.471\,992\,85 \times 10^{-2}$, $-2.000\,000\,00 \times 10^{-2}$, $-1.341\,583\,10 \times 10^{-4}$, $1.341\,583\,10 \times 10^{-4}$, $2.000\,000\,00 \times 10^{-2}$, $4.471\,992\,85 \times 10^{-2}$, $1.967\,857\,12$, $1.987\,7\,229\,6$, $2.012\,442\,89$, $2.032\,577\,04$, $4.000\,300\,00$.

Comparing with H_T, one can see the effects of order J and J^2. It is a good exercise to reproduce these results using degenerate perturbation theory.

Exercise 6: Calculate $|\langle 1000|U(t)|1000\rangle|^2$ with $U(t) = \exp(-itH)$ for t between 0 and 1000 (Jt between 0 and 20) Note: $U(t) = \sum_n \exp(-iE_n t)|n\rangle\langle n|$, with $H|n\rangle = E_n|n\rangle$. Display the evolution graphically and compare with the same quantity with the Trotter procedure $(\exp(\epsilon(\hat{A} + \hat{B})) \simeq \exp(\epsilon\hat{A})\exp(\epsilon\hat{B}) + \mathcal{O}(\epsilon^2))$ for $\delta t = 5$, 1, and 0.5

Solution. See figures 9.8 and 9.9.

Exercise 7: Using PBC compare $|\langle 1000|U(t)|1000\rangle|^2$ with $|J_0^{(N_s)}(2Jt)|^2$ for t between 0 and 1000.

Solution. See figures 9.10 and 9.11.

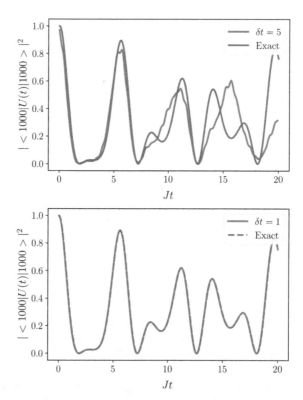

Figure 9.8. $|\langle 1000|U(t)|1000\rangle|^2$ versus Jt for $h_T = 1$ and $J = 0.02$ OBC using exact diagonalization and the Trotter approximation for $\delta t = 5$ (top) and $\delta 1 = 1$ (bottom).

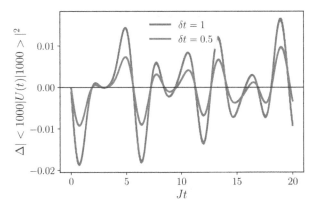

Figure 9.9. $\Delta|\langle1000|U(t)|1000\rangle|^2$ versus Jt for $h_T = 1$ and $J = 0.02$ OBC: difference between exact diagonalization and the Trotter approximation for $\delta t = 1$ and $\delta t = 0.5$.

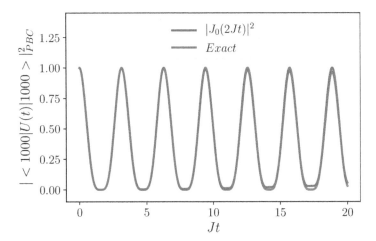

Figure 9.10. $|\langle1000|U(t)|1000\rangle|^2$ versus Jt for $h_T = 1$ and $J = 0.02$ PBC: exact diagonalization and the Bessel approximation.

9.7 Rigorous and empirical Trotter bounds

As discussed in section 5.4, for finite matrices [44] $S_n = e^{\frac{A+B}{n}}$ and $T_n = e^{\frac{A}{n}}e^{\frac{B}{n}}$, we can find a bound on the error by using the one step error:

$$||S_n^n - T_n^n|| \leqslant n(\max(||S_n||, ||T_n||))^{n-1} ||S_n - T_n||. \tag{9.65}$$

Since the error on one step is of second order, we expect

$$||S_n - T_n|| \leqslant \frac{C}{n^2}, \tag{9.66}$$

and $||S_n^n - T_n^n||$ can be made as small as we want by increasing n. The constant of proportionality can be estimated perturbatively by expanding in powers of ϵ

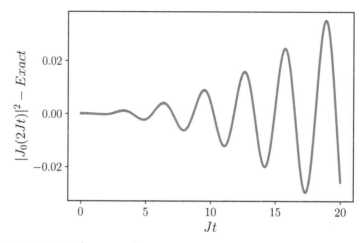

Figure 9.11. $|\langle 1000|U(t)|1000\rangle|^2$ versus Jt for $h_T = 1$ and $J = 0.02$ PBC: difference between exact diagonalization and the Bessel approximation.

$$e^{\epsilon A}e^{\epsilon B} - e^{\epsilon(A+B)} \simeq -\frac{\epsilon^2}{2}[A, B]. \tag{9.67}$$

A longer calculation leads to the higher order bound

$$e^{\epsilon A/2}e^{\epsilon B}e^{\epsilon A/2} - e^{\epsilon(A+B)} \simeq \frac{\epsilon^3}{24}(2[B[B, A]] - [A[A, B]]). \tag{9.68}$$

For practical implementations with quantum computers, the number of Trotter steps n may be limited by coherence and noise issues. In this context, it should be emphasized that equation (9.65) is an upper bound and that in specific situations it may not have been saturated. For similar reasons, it should be recognized that the quadratic or cubic behavior in the time step may require small time steps but that for larger time steps the one step error may grow at a smaller rate.

These nonlinear aspects will be studied with the quantum Ising model with four sites and OBC. For a matrix M, we will use the norm $||M|| = \sqrt{\lambda}$ with λ the largest eigenvalue of $M^\dagger M$. For numerical purpose, we will focus on the case $h_T = 1$ and $J = 0.02$ studied in reference [42, 45]

$$\frac{A}{n} \rightarrow -i\hat{H}_T h_T \delta t, \tag{9.69}$$

$$\hat{H}_T = -\sum_{j=1}^{4} \hat{\sigma}_j^z, \tag{9.70}$$

and

$$\frac{B}{n} \rightarrow -i\hat{H}_{NN}J\delta t, \tag{9.71}$$

$$\hat{H}_{NN} = -\sum_{j=1}^{3} \hat{\sigma}_j^x \hat{\sigma}_{j+1}^x. \tag{9.72}$$

We first consider the one-step discrepancy for the simplest Trotter approximation where the discrepancy is of order $(\delta t)^2$

$$\Delta_2 U \equiv e^{-i(h_T \hat{H}_T + J \hat{H}_{NN})\delta t} - e^{-ih_T \hat{H}_T \delta t} e^{-iJ \hat{H}_{NN} \delta t}$$

$$\simeq \frac{h_T J}{2} [\hat{H}_T, \hat{H}_{NN}](\delta t)^2. \tag{9.73}$$

Empirically, we find that the bound

$$\|\Delta_2 U\| \leqslant \frac{h_T J}{2} \| [\hat{H}_T, \hat{H}_{NN}] \| (\delta t)^2, \tag{9.74}$$

is sharp for δt small enough. Numerically we find that for $N_s = 4$.

$$\| [\hat{H}_T, \hat{H}_{NN}] \| \simeq 8.944. \tag{9.75}$$

The results for $h_T = 1$ and $J = 0.02$ are displayed in figure 9.12. The bound is sharp for $\delta t \leqslant 0.3$.

We have repeated the comparison for the more accurate Trotter approximation where we have

$$\Delta_3 U \equiv e^{-i(h_T \hat{H}_T + J \hat{H}_{NN})\delta t} - e^{-ih_T \hat{H}_T \delta t/2} e^{-iJ \hat{H}_{NN} \delta t} e^{-ih_T \hat{H}_T \delta t/2}, \tag{9.76}$$

$$\simeq \left(\frac{-h_T^2 J}{24} [\hat{H}_T, [\hat{H}_T, \hat{H}_{NN}]] + \frac{h_T J^2}{12} [\hat{H}_{NN}, [\hat{H}_T, \hat{H}_{NN}]] \right)(\delta t)^3. \tag{9.77}$$

We now consider the bound.

$$\|\Delta_3 U\| \leqslant \left(\frac{h_T^2 J}{24} \| [\hat{H}_T, [\hat{H}_T, \hat{H}_{NN}]] \| + \frac{h_T J^2}{12} \| [\hat{H}_{NN}, [\hat{H}_T, \hat{H}_{NN}]] \| \right)(\delta t)^3. \tag{9.78}$$

Numerically

$$\| [\hat{H}_T, [\hat{H}_T, \hat{H}_{NN}]] \| \simeq 35.777, \tag{9.79}$$

and

$$\| [\hat{H}_{NN}, [\hat{H}_T, \hat{H}_{NN}]] \| \simeq 28.844. \tag{9.80}$$

The results for $h_T = 1$ and $J = 0.02$ are displayed in figure 9.13. The bound is sharp for $\delta t \leqslant 0.7$. However, for larger δt, the two actual one-step errors cross around 1.5 as shown in figure 9.14. For larger δt, the less accurate approximation (at small δt) become somehow more accurate as shown in figures 9.15 and 9.16. In addition, the lowest order $\|\Delta_2 U\|$ has bumps at $\delta t = \pi/2, 3\pi/4, \pi \cdots$. It is easy to see that $\delta t = \pi$ is very special. For $h_T = 1$, the spectrum of $\hat{H}_T$ has even eigenvalues and for $\delta t = \pi$, $e^{-ih_T \hat{H}_T \delta t}$ reduces to

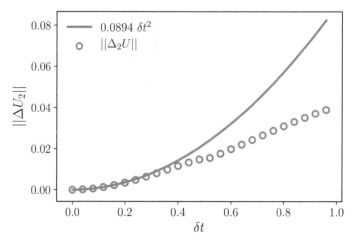

Figure 9.12. $\|\Delta_2 U\|$ versus δt for $h_T = 1$ and $J = 0.02$.

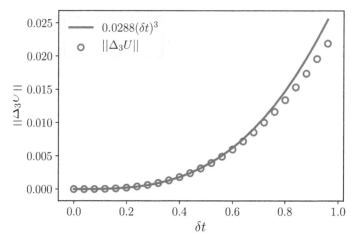

Figure 9.13. $\|\Delta_2 U\|$ versus δt for $h_T = 1$ and $J = 0.02$.

the identity. Consequently, for this very special value of δt, the leading term of the Hamiltonian completely disappears from the Trotter approximation!

9.8 Optimal Trotter error

The choice of an optimal Trotter step is difficult. From a rigorous point of view, it is clear that we should have steps as small as possible. From a practical point view, we have a very limited number of steps at our disposal if we use a quantum computer. With these conflicting requirements, the optimal choice will depend on the type of observables used, the choices of couplings and physical properties that we are interested in.

One of our main objectives is to study real time scattering processes where wave packets are prepared at distances where their mutual interactions are negligible and then evolved in such a way that scattering processes take place. The time relevant

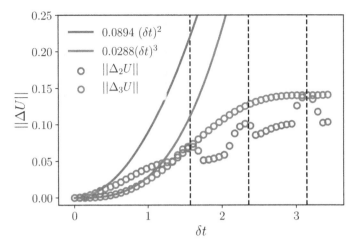

Figure 9.14. $||\Delta_2 U||$ and $||\Delta_3 U||$ versus δt for $h_T = 1$ and $J = 0.02$. The vertical lines are at $\pi/2$, $3\pi/4$ and π.

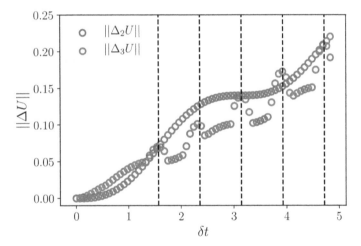

Figure 9.15. $||\Delta_2 U||$ and $||\Delta_3 U||$ versus δt for $h_T = 1$ and $J = 0.02$. The vertical lines are at $\pi/2$, $3\pi/4$ and $\pi \cdots$.

time scale is an intermediate one in between the scale of the short time propagation and the long time scale relevant to study equilibration of a set of particles.

Our choice of parameters $h_T = 1$ and $J = 0.02$ provides a large gap between the vacuum and the one-particle states with small splittings that can be interpreted as kinetic energy. From the point of view of calculating the spectrum J is a perturbation, but for the real-time evolution of the $\langle n_j(t) \rangle$ with initial states which are eigenstates of n_j, this quantity remains constant in the limit $J = 0$. Consequently, the changes in $\langle n_j(t) \rangle$ are driven by J. For this observable, the rapid oscillations associated with $\hat{H}_T$ have no effects if applied alone, while $\hat{H}_{NN}$ creates slow changes.

The time scale associated with $\hat{H}_{NN}$ can be estimated by looking at its spectrum. For $N_s = 4$ with OBC and $J = 0.02$, the ground state is doubly degenerate with an energy -0.06. The entire spectrum can be calculating as the energy of a one-dimensional

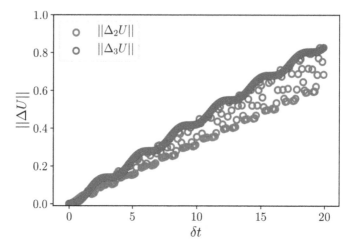

Figure 9.16. $||\Delta_2 U||$ and $||\Delta_3 U||$ versus δt for $h_T = 1$ and $J = 0.02$.

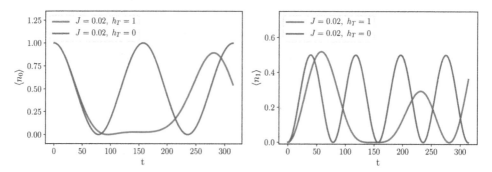

Figure 9.17. $\langle n_0(t) \rangle$ (left) and $\langle n_1(t) \rangle$ (right) with an initial state $|1000\rangle$ for $h_T = 1$ and $J = 0.02$ (red) and $h_T = 0$ and $J = 0.02$.

classical Ising model. The other energies are -0.02, 0.02 and 0.06 with respective multiplicities 6, 6 and 2. Consequently, the evolution operator associated with $\hat{H}_{NN}$ has a periodicity $2\pi/0.02 \simeq 314$. It is actually interesting to consider the effect of turning on h_T from 0 to 1 with $J = 0.02$ kept constant. This is illustrated in figure 9.17. We see that for t up to 50, the effects of the leading term H_T are somehow mild.

If we look at the exact evolution with $h_T = 1$ and $J = 0.02$, we see that substantial changes occur after $t = 100$. With currently available quantum computers, a choice of δt between 10 and 20 could be considered as a good choice because between 10 and 5 steps would reach $t = 100$. For such values of δt, the rigorous bounds proportional to $(\delta t)^2$ and $(\delta t)^3$ discussed above are not useful because they are only valid for $\delta t \lesssim 1$. This is a nonlinear problem and as a first approach we can compare the numerical results with the exact evolution and the Trotter approximation for various δt. This is done in figures 9.18–9.21. We see that $\delta t = 10$ stays very close to the exact results with fluctuations slightly larger for low occupations, but much larger discrepancies are

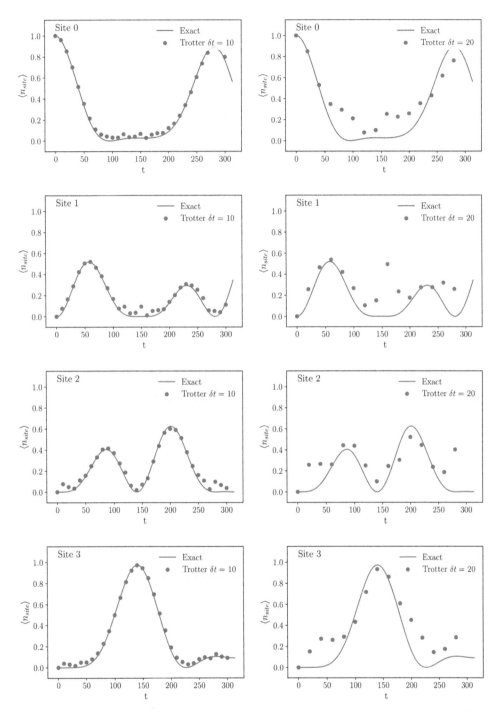

Figure 9.18. $\langle n_0(t) \rangle$, $\langle n_1(t) \rangle$, $\langle n_2(t) \rangle$ and $\langle n_3(t) \rangle$ (from top to bottom) for $\delta t = 10$ (left) and 20 (right) with an initial state $|1000\rangle$ for $h_T = 1$ and $J = 0.02$. The exact evolution is shown in red and the Trotter approximation in blue.

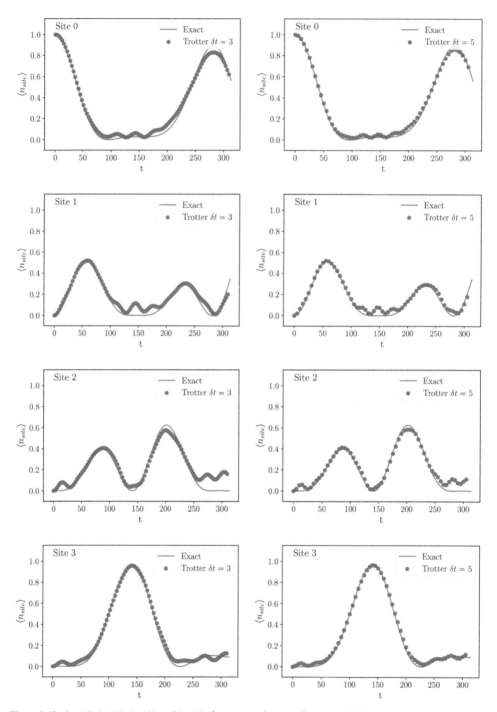

Figure 9.19. $\langle n_0(t) \rangle$, $\langle n_1(t) \rangle$, $\langle n_2(t) \rangle$ and $\langle n_3(t) \rangle$ (from top to bottom) for $\delta t = 3$ (left) and 5 (right) with an initial state $|1000\rangle$ for $h_T = 1$ and $J = 0.02$. The exact evolution is shown in red and the Trotter approximation in blue.

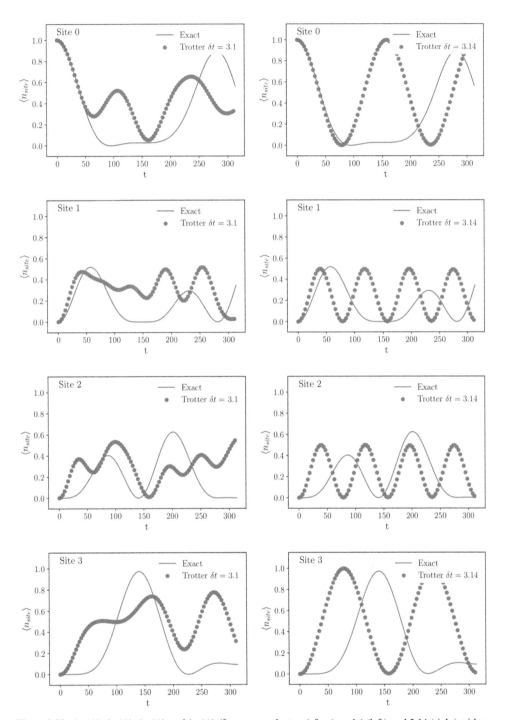

Figure 9.20. $\langle n_0(t) \rangle$, $\langle n_1(t) \rangle$, $\langle n_2(t) \rangle$ and $\langle n_3(t) \rangle$ (from top to bottom) for $\delta t = 3.1$ (left) and 3.14 (right) with an initial state $|1000\rangle$ for $h_T = 1$ and $J = 0.02$. The exact evolution is shown in red and the Trotter approximation in blue.

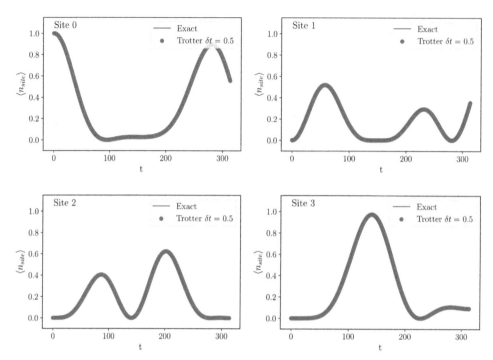

Figure 9.21. $\langle n_0(t) \rangle, \langle n_1(t) \rangle, \langle n_2(t) \rangle$ and $\langle n_3(t) \rangle$ (from top to bottom) for $\delta t = 0.5$ with an initial state $|1000\rangle$ for $h_T = 1$ and $J = 0.02$. The exact evolution is shown in red and the Trotter approximation in blue.

observed for $\delta t = 20$. Reducing further to $\delta t = 5$ improves slightly the accuracy, however, $\delta t = 3$ does not appear to be an improvement. Looking at other values near 3 such as $\delta t = 3.1$ or 3.14 creates larger discrepancies. This has a simple explanation: $\hat{H}_T$ has eigenvalues ± 4, ± 2 and 0 and the associated evolution has a π periodicity. Consequently, for $\delta t = n\pi$, $\hat{H}_T$ does not appear at all in the Trotter approximation. If we reduce δt to 0.5, then we are in the regime where the $(\delta t)^2$ and $(\delta t)^3$ bounds are valid and the Trotter approximation is excellent but requires 200 steps to reach $t = 100$. These results are illustrated in the figures 9.18–9.21.

A more comprehensive list of references can be found in: Meurice Y, Sakai R and Unmuth-Yockey J Tensor field theory with applications to quantum computing arXiv:2010.06539.

References

[1] Wilson K G and Kogut J B 1974 *Phys. Rep.* **12** 75–199

[2] Fradkin E H and Susskind L 1978 *Phys. Rev.* D **17** 2637

[3] Kogut J B 1979 *Rev. Mod. Phys.* **51** 659–713

[4] Luscher 1990 Selected topics in lattice field theory *Champs, Cordes, Et Phénomènes Critiques, Les Houches 1988, Session 49* ed E Brézin and J Zinn-Justin (Amsterdam: North-Holland)

[5] Zou H, Liu Y, Lai C Y, Unmuth-Yockey J, Bazavov A, Xie Z Y, Xiang T, Chandrasekharan S, Tsai S W and Meurice Y 2014 *Phys. Rev.* A **90** 063603

[6] Banks T, Susskind L and Kogut J 1976 *Phys. Rev.* D **13** 1043–53

[7] Meurice Y 2019 *Phys. Rev.* D **100** 014506

[8] Bruckmann F, Jansen K and Kühn S 2019 *Phys. Rev.* D **99** 074501

[9] Brower R, Chandrasekharan S and Wiese U J 1999 *Phys. Rev.* D **60** 094502

[10] Vilenkin N 1978 *Special Functions and the Theory of Group Representations, Translations of Mathematical Monographs* (Providence, RI: American Mathematical Society)

[11] Kadoh D, Kuramashi Y, Nakamura Y, Sakai R, Takeda S and Yoshimura Y 2018 *J. High Energy Phys.* **03** 141

[12] Kadoh D, Kuramashi Y, Nakamura Y, Sakai R, Takeda S and Yoshimura Y 2019 *J. High Energy Phys.* **05** 184

[13] Kadoh D, Kuramashi Y, Nakamura Y, Sakai R, Takeda S and Yoshimura Y 2020 2020 *J. High Energy Phys.* **02** 161

[14] Bazavov A, Meurice Y, Tsai S W, Unmuth-Yockey J and Zhang J 2015 *Phys. Rev.* D **92** 076003

[15] Meurice Y 2020 *Phys. Rev.* D **102** 014506

[16] Unmuth-Yockey J F 2019 *Phys. Rev.* D **99** 074502

[17] Tagliacozzo L, Celi A and Lewenstein M 2014 *Phys. Rev.* X **4** 041024

[18] Haegeman J, Van Acoleyen K, Schuch N, Cirac J I and Verstraete F 2015 *Phys. Rev.* X **5** 011024

[19] Savit R 1980 *Rev. Mod. Phys.* **52** 453–87

[20] Unmuth-Yockey J F 2017 Duality methods and the tensor renormalization group: applications to quantum simulation *PhD Thesis* University of Iowa https://ir.uiowa.edu/etd/5869/

[21] Kaplan D B and Stryker J R 2018 (arXiv:1806.08797)

[22] Bender J and Zohar E 2008 (arXiv:2008.01349)

[23] Jackson J D 1998 *Classical Electrodynamics* (New York: Wiley)

[24] Fannes M, Nachtergaele B and Werner R F 1992 *Commun. Math. Phys.* **144** 443–90

[25] Vidal G 2003 *Phys. Rev. Lett.* **91** 147902

[26] Vidal G 2004 *Phys. Rev. Lett.* **93** 040502

[27] Verstraete F and Cirac J I 2004 (arXiv:cond-mat/0407066)

[28] Shi Y Y, Duan L M and Vidal G 2006 *Phys. Rev.* A **74** 022320

[29] Perez-Garcia D, Verstraete F, Wolf M M and Cirac J I 2006 *Matrix product state representations* (arXiv:quant-ph/0608197)

[30] Verstraete F, Murg V and Cirac J 2008 *Adv. Phys.* **57** 143–224

[31] White S R 1992 *Phys. Rev. Lett.* **69** 2863–6

[32] Schollwock U 2005 *Rev. Mod. Phys.* **77** 259–315

[33] Levin M and Nave C P 2007 *Phys. Rev. Lett.* **99** 120601

[34] Gu Z C and Wen X G 2009 *Phys. Rev.* B **80** 155131

[35] Efrati E, Wang Z, Kolan A and Kadanoff L P 2014 *Rev. Mod. Phys.* **86** 647–67

[36] Xie Z Y, Chen J, Qin M P, Zhu J W, Yang L P and Xiang T 2012 *Phys. Rev.* B **86** 045139

[37] Meurice Y 2013 *Phys. Rev.* B **87** 064422

[38] Migdal A A 1975 *Sov. Phys. JETP* **42** 743 http://www.jetp.ac.ru/cgi-bin/dn/e_042_04_0743.pdf

Migdal A A 1975 *Zh. Eksp. Teor. Fiz.* **69** 1457

[39] Kadanoff L P 1976 *Ann. Phys.* **100** 359–94

[40] Itzykson C and Drouffe J 1992 *Statistical Field Theory, Cambridge Monographs on Mathematical Physics* (Cambridge: Cambridge University Press)

[41] Liu Y, Meurice Y, Qin M P, Unmuth-Yockey J, Xiang T, Xie Z Y, Yu J F and Zou H 2013 *Phys. Rev.* D **88** 056005

[42] Gustafson E, Meurice Y and Unmuth-Yockey J 2019 *Phys. Rev.* D **99** 094503
[43] Lloyd S 1996 *Science* **273** 1073–8
[44] Reed M and Simon B 1981 *I: Functional Analysis, Methods of Modern Mathematical Physics* (Amsterdam: Elsevier)
[45] Gustafson E, Dreher P, Hang Z and Meurice Y 2019 (arXiv:1910.09478)

IOP Publishing

Quantum Field Theory
A quantum computation approach
Yannick Meurice

Chapter 10

Recent progress in quantum computation/ simulation for field theory

10.1 Analog simulations with cold atoms

One very important contribution from the atomic and molecular physics community to 'analog quantum computing' is the development of quantum simulations experiments recreating theoretical lattice models in a volume smaller than 1 mm^3. We could call these experimental setups 'optical lattices computers'. A review of the birth of this new field can be found in reference [1]. Alkali-metals such as Li, Na, K, Rb, and Cs have a loosely bound electron in the outer shell and develop a dipole moment when placed in an electric field. Typical choices are ^{87}Rb (a boson: 37 e$^-$, 37 p and 50 n) or ^{6}Li (a fermion: 3 e$^-$, 3 p and 3 n). Polarizable cold atoms trapped in standing waves created by counterpropagating laser beams in 1, 2 or 3 dimensions will generate a periodic potential due to the dipole moment induced by the linearly polarized laser beam:

$$V(\mathbf{r}) = -(1/2)\alpha(\omega)|\mathbf{E}(\mathbf{r})|^2, \tag{10.1}$$

with

$$\alpha(\omega) \sim |\langle e|d|g\rangle|^2/\hbar(\omega_0 - \omega_L), \tag{10.2}$$

where ω_0 and ω_L are the angular frequency corresponding to the dipole resonance and the laser frequency, respectively.

A three-dimensional optical lattice potential with a cubic symmetry can be created using three mutually orthogonal laser beams of the same wavelength λ_L. The periodic potential is

$$V(x, y, z) = V_0(\sin^2(kx) + \sin^2(ky) + \sin^2(kx)), \tag{10.3}$$

with $k = 2\pi/\lambda_L$. The lattice spacing is $a = \lambda_L/2$. The depth of the potential V_0 is measured in units of the recoil energy $E_r \equiv (\hbar k)^2/2\, m_{\text{atom}}$ and can be tuned

doi:10.1088/978-0-7503-2187-7ch10

continuously by changing the intensity of the laser. For rubidium atoms with $\lambda_L = 856$ nm, the recoil energy is 1.3×10^{-11} eV $\simeq k_B 1.5 \times 10^{-7}$ K. The critical temperature for Bose condensation in rubidium with a specific volume of $(\lambda_L/2)^3$ is close to 10^{-7} K according to the ideal gas formula. The recoil momentum is 1.5 eV c^{-1} and the recoil velocity about 5 mm s^{-1}. Of the order of $N_{\text{atoms}} \simeq 65^3$ were used in early experiments. Assuming one atom per site, the physical size of the lattice is of the order of 30 μm.

The momentum distribution of a many-body quantum state can be measured by turning off the optical lattice and taking pictures of the density at various times. It takes a few milliseconds for atoms moving at speeds of the order of the recoil velocity to go across the lattice size. When the depth of the potential V_0 is increased up to 20 E_r, the harmonic frequency for the atoms in the periodic potential is approximately 30 kHz. The insulator to superfluid transition for the Bose–Hubbard model was studied with this type of methods [1]. For this simple model, quantum Monte Carlo methods are very reliable and the agreement between theory and experiments for the time-of-flight pictures is remarkable [2]. Optical lattice experiments fit on average size tables but they can be quite complex. Examples are shown in figure 10.1.

Optical lattice experiments can also use fermions. A very interesting model to be quantum simulated is the the Fermi–Hubbard model. Its Hamiltonian is

$$H = -t \sum_{\langle \mathbf{x},\mathbf{y} \rangle, \alpha} \left(f^{\dagger}_{\mathbf{x},\alpha} f_{\mathbf{y},\alpha} + h.c. \right) + U \sum_{\mathbf{x}} n_{\mathbf{x}\uparrow} n_{\mathbf{x}\downarrow}$$

where t characterizes the tunneling between nearest neighbor sites and U controls the onsite Coulomb repulsion. The $f^{\dagger}_{\mathbf{x},\alpha}$ and $f_{\mathbf{x},\alpha}$ obey canonical anticommutation relations and the spin degree of freedom $\alpha = \uparrow$ or $\downarrow$ is necessary to have two fermions at the same site. The interactions are illustrated in figure 10.2. The Bose–Hubbard model has similar type of interactions but rely on commutation relations and so the spin degree of freedom is not necessary to have onsite interactions.

Figure 10.1. An optical lattice experiment in Immanuel Bloch's group (Garching, Munich, left, picture from Axel Griesch) and Cheng Chin's group (right, picture from U. Chicago). Courtesy of I Bloch and C Chin.

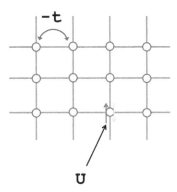

Figure 10.2. Graphical representation of the interactions of the Hubbard model. Reprinted with permission of APS from [3].

All these interactions can be approximately recreated with atoms trapped in an optical lattice.

In the strong coupling limit ($U \gg t$) and at half-filling (one fermion per site), the Fermi–Hubbard is approximately described by quantum Heisenberg model

$$H = J \sum_{\langle xy \rangle} \mathbf{S_x} \cdot \mathbf{S_y}. \tag{10.4}$$

The coupling $J = 4t^2/U$ as follows from second-order degenerate perturbation theory for the Hubbard model. We use the definition

$$\mathbf{S_x} \equiv \frac{1}{2} f^{\dagger}_{x\alpha} \sigma_{\alpha\beta} f_{x\beta}. \tag{10.5}$$

One can check that the anticommutations relations imply that

$$\left[\mathbf{S}^j_x, \mathbf{S}^k_y \right] = i\epsilon^{jkl} \mathbf{S}^l_x \delta_{x,y}. \tag{10.6}$$

Using the Fierz identity

$$\vec{\sigma}_{\alpha\beta} \cdot \vec{\sigma}_{\gamma\delta} = 2\delta_{\alpha\delta}\delta_{\beta\gamma} - \delta_{\alpha\beta}\delta_{\gamma\delta}, \tag{10.7}$$

and the anticommutations relations, the Heisenberg Hamiltonian becomes

$$H = -\sum_{\langle xy \rangle} \frac{1}{2} J f^{\dagger}_{x\alpha} f_{y\alpha} f^{\dagger}_{y\beta} f_{x\beta} + \sum_{\langle xy \rangle} J\left(\frac{1}{2} n_x - \frac{1}{4} n_x n_y \right), \tag{10.8}$$

with

$$n_x \equiv f^{\dagger}_{x\alpha} f_{x\alpha}, \tag{10.9}$$

which takes the values 0, 1, and 2. The first term corresponds to second order perturbation theory for the Fermi–Hubbard model. Note that

$$\mathbf{S_x} \cdot \mathbf{S_x} = \frac{3}{4} n_x (2 - n_x), \tag{10.10}$$

is only non-zero for $n_x = 1$ which is imposed in order to recover the Heisenberg model.

The Heisenberg reformulation has a hidden local invariance [4] that we proceed to discuss with slightly different notations. Note that

$$\begin{pmatrix} f_1 \\ f_2 \end{pmatrix} \text{ and } \begin{pmatrix} f_2^\dagger \\ -f_1^\dagger \end{pmatrix}, \tag{10.11}$$

transform identically under a global $SU(2)$ rotation U. This comes from the identity $U^\star = \sigma^2 U \sigma^2$. Since the rotation is global, we omit the x index. We can combine these two doublets in a 2×2 matrix

$$M \equiv \begin{pmatrix} f_1 & f_2^\dagger \\ f_2 & -f_1^\dagger \end{pmatrix}. \tag{10.12}$$

A short calculation shows that

$$-MM^\dagger = \vec{S} \cdot \vec{\sigma} - \mathbb{1}, \tag{10.13}$$

and consequently

$$\vec{S} = -\frac{1}{2}\text{Tr}(MM^\dagger \vec{\sigma}.). \tag{10.14}$$

Under a global rotation

$$MM^\dagger \rightarrow UMM^\dagger U^\dagger, \tag{10.15}$$

which within the trace implies

$$\vec{\sigma} \rightarrow U^\dagger \vec{\sigma} U. \tag{10.16}$$

This amounts to an inverse rotation of the three-vector $\vec{\sigma}$ and $\mathbf{S}$. Since the rotation is global $\mathbf{S_x} \cdot \mathbf{S_y}$ is invariant.

In addition, there is a *local* transformation which leaves $M_x M_x^\dagger$ and $\mathbf{S_x}$ invariant:

$$M_x \rightarrow M_x V_x^\dagger, \text{ and } M_x^\dagger \rightarrow V_x M_x^\dagger. \tag{10.17}$$

The basic doublets of this local symmetry are (omitting the local label x).

$$\begin{pmatrix} f_1^\dagger \\ f_2 \end{pmatrix} \text{ and } \begin{pmatrix} f_2^\dagger \\ -f_1 \end{pmatrix}, \tag{10.18}$$

It is possible to use a particle–hole transformation say for f_1: $\tilde{f}_1 = f_1^\dagger$ and $\tilde{f}_2 = f_2$, and the local doublets look like the global doublets. The local invariant quantities $\tilde{f}_{x\alpha}^\dagger \tilde{f}_{x\alpha}$ and $\epsilon^{\alpha\beta} \tilde{f}_{x\alpha} \tilde{f}_{x\beta}$ can be interpreted as meson and $SU(2)$ baryons as they appear in the strong coupling expansion of gauge theories [5]. Proposals for quantum simulations are discussed in reference [3].

Early proposals to quantum simulate lattice gauge theories with optical lattices are discussed in [6]. More references can be found in review articles [7–10]. For recent experimental developments see [11–14]. Trapped ions [15] (see figure 10.3) provide controllable all-to-all connectivity and open new opportunities to approach lattice gauge theory models [16]. Rydberg atoms offer a versatile platform to study many-body problems [17], and out-of-equilibrium properties of spin models [18] and gauge theories [19]. The Schwinger model is often the first target when developing new approaches [12, 16, 20–24]. For recent work on non-abelian models see [25–29]. For a recent articles reviewing various simulation methods with a focus on lattice gauge theories see [10, 30].

10.2 Experimental measurement of the entanglement entropy

The concept of entanglement entropy is very important to understand the critical behavior of lattice models. Remarkably, it is possible to measure it experimentally [31, 32] for models discussed in the previous section. In order to define the entanglement entropy, we consider the subdivision of a system AB into subsystems A and B. We define the reduced density matrix $\hat{\rho}_A$ as

$$\hat{\rho}_A \equiv \text{Tr}_B \hat{\rho}_{AB}, \tag{10.19}$$

and the von Neumann entanglement entropy as

$$S_{\text{EvonNeumann}} = -\sum_i \rho_{A_i} \ln(\rho_{A_i}). \tag{10.20}$$

Figure 10.3. Guido Pagano and a trapped ion experiment at University of Maryland.

It is also interesting to introduce the nth order Rényi entanglement entropy:

$$S_n(A) \equiv \frac{1}{1-n} \ln(\text{Tr}((\hat{\rho}_A)^n)). \tag{10.21}$$

The von Neumann entanglement entropy can be defined as the $\lim_{n \to 1^+} S_n$.

The basic idea behind the experimental measurement of the second-order Rényi entanglement entropy is to swap two copies of a given system in order to measure $\text{Tr}(\rho_A^2)$:

$$\exp(-S_2) = \text{Tr}(\rho_A^2) = \text{Tr}(\text{Swap}(\rho_{A1} \otimes \rho_{A2})) \tag{10.22}$$

Experimentalists can create one-dimensional 'tubes' of cold atoms trapped in an optical lattice and put twin copies of these tubes side-by-side. Following [33], one can use tunneling between the two copies that produces a beam-slitting operation such that

$$\text{Tr}(\text{Swap}(\rho_{A1} \otimes \rho_{A2})) = \left\langle (-1)^{\sum_{x \in A} n_x^{\text{copy}}} \right\rangle. \tag{10.23}$$

After the beamsplitter operation, one selects one copy of the twin tubes and measure the number of particles modulo 2 at each site x of this copy, denoted n_x^{copy}. This method was successfully applied with ^{87}Rb atoms [31] which can be described by the Bose–Hubbard Hamiltonian:

$$H = \frac{U}{2} \sum_x n_x(n_x - 1) - J \sum_x (a_x^\dagger a_{x+1} + h.c.).$$

The entanglement entropy of the Bose–Hubbard (BH) Hamiltonian is closely related to the entanglement entropy of the $O(2)$ model with a sufficiently large chemical potential as introduced in equation (9.3). In both cases, there are superfluid regions in the phase diagram where in $1 + 1$ dimensions, the models can be described by conformal field theory (CFT) methods for a central charge $c = 1$. For an introduction to these methods and the meaning of the central charge see [34]. 2D CFT predicts [35–38] that for a system of spatial size N_s, S_n has an approximately linear behavior in $\ln(N_s)$. This is called the Calabrese–Cardy scaling. For a central charge c, it has the form

$$S_n(N_s) = K + \frac{c(n+1)}{6n} \ln(N_s),$$

for periodic boundary conditions and half the slope $(\frac{c(n+1)}{12n})$ for open boundary conditions. Figure 10.4 shows an explicit calculation for the two models discussed in reference [39]. A three-state approximation was used for the $O(2)$ model. The constant A (predicted to be 0.125) is fitted using the parametrization [37, 38, 40, 41] for subsystems of size $N_s/2$:

$$S_2(N_s) = K + A \ln(N_s) + \frac{B \cos\left(\frac{\pi N_s}{2}\right)}{(N_s)^p} + \frac{D}{\ln^2(N_s)}. \tag{10.24}$$

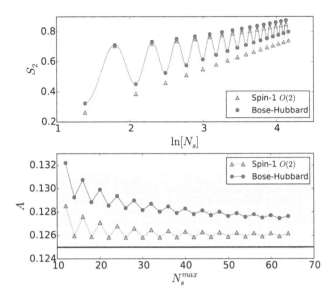

Figure 10.4. (Top) S_2 at half-filling with OBC for $O(2)$ and BH with $J/U = 0.1$. The solid lines are the fits for BH and $O(2)$. (Bottom) values of A as a function of the maximal value of N_s used in the fit, the band represents a positive departure of 5 percent from the expected value 0.125. Reprinted with permission of APS from [39].

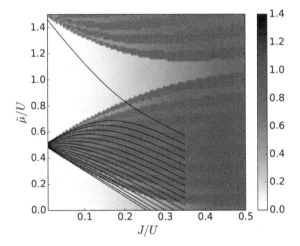

Figure 10.5. S_2 for $O(2)$ with $N_s = 16$ and OBC. Laid over top are the BH boundaries between particle number sectors. Reprinted with permission of APS from [39].

The fit for A is reasonably close to the predicted value 0.125. The values of S_2 for $O(2)$ as a function of the chemical potential and hopping parameter are shown in figure 10.5. See [39, 42] for a complete discussion of the parameters.

10.3 Implementation of the abelian Higgs model

The tensor formulation and the time continuum limit for the compact abelian Higgs model have been discussed in sections 6.5, 7.6 and 9.2. In the following we concentrate on 1 + 1 dimensions where there is no space–space plaquette. Using notations closer to references [43, 44], where the electric field is denoted $\bar{L}^z$, the Hamiltonian of equation (9.37) with colors corresponding to figure 10.6 reads

$$\bar{H} = \frac{\tilde{U}_g}{2} \sum_i \left(\bar{L}^z_{(i)}\right)^2 + \frac{\tilde{Y}}{2} \sum_i \left(\bar{L}^z_{(i)} - \bar{L}^z_{(i+1)}\right)^2 - \tilde{X} \sum_i \bar{U}_{(i)}. \tag{10.25}$$

The vertical dimension corresponds to the quantum number associated with the electric field which takes the values -2, -1, 0, 1, 2 after what we call the spin-2 truncation. The horizontal dimension corresponds to the spatial dimension with nine lattice sites. This ladder structure can be recreated with Rydberg atoms on an optical lattice [43]. There is only one atom per rung and tunneling occurs along the vertical direction (green). There is no tunneling in the horizontal direction but short range attractive interactions (blue). A parabolic potential is applied in the spin (vertical) direction to recreate the first term of the Hamiltonian (onsite energy, red). The main difficulty is to implement the nearest neighbor interactions (blue) with a Rydberg potential. This can be accomplished by using an asymmetric lattice where the spatial spacing is much larger than the spin spacing and we can use perturbation in the short side of the Pythagoras expression to get the quadratic (blue) term when the spatial

5 states ladder with 9 rungs

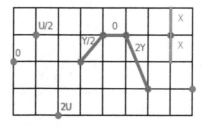

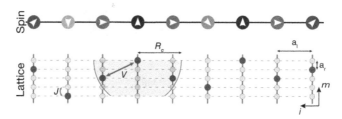

Figure 10.6. Multi-leg ladder implementation discussed in the text. Top: a five-leg ladder with nine rungs used to simulate the Hamiltonian (10.25) with a spin-2 truncation and $N_s = 9$. Bottom: experimental implementation [43]. The spin part shows the possible m_z-projections. Below, we show the corresponding realization in a ladder within an optical lattice. The atoms (green disks) are allowed to hop within a rung with a strength J, while no hopping is allowed along the legs. The lattice constants along rung and legs are a_r and a_l, respectively. Coupling between atoms in different rungs is implemented via an isotropic Rydberg-dressed interaction V with a cutoff distance R_c (marked by blue shading). For details see [43]. Reprinted with permission of APS from [43].

spacing is close to the distance where the Rydberg potential changes substantially. This is illustrated in figure 10.6 and explained in more detail in reference [43].

The quantity that we propose to measure is the Polyakov loop, a Wilson line wrapping around the Euclidean time direction:

$$\langle P_i \rangle = \left\langle \prod_j U_{(i,j),\tau} \right\rangle = \exp(-F(\text{single charge})/kT), \qquad (10.26)$$

with the order parameter for the deconfinement transition at finite temperature in the pure gauge case [45]. With periodic boundary condition, the insertion of the Polyakov loop (red) forces the presence of a scalar current (green) in the opposite direction (left) or another Polyakov loop (right) (figure 10.7). In the Hamiltonian formulation, we can insert a Polyakov loop by adding $-\frac{\bar{Y}}{2}(2(\bar{L}_i^z - \bar{L}_{(i+1)}^z) - 1)$ to H with i the location of the Polyakov loop. This quantity satisfies what we call data collapse (figure 10.8) when we consider data from different volumes with a suitable rescaling. This is called finite size scaling and it allows us to exploit data from small size systems. The general concept will be explained in section 11.6.

10.4 A two-leg ladder as an idealized quantum computer

We can develop some intuition about the ladder setup by considering the idealized case of a two-leg ladder emulating the quantum Ising model (spin-1/2) with a Hamiltonian given in equation (9.53). In the experimental setup, there is exactly one atom per rung and we can identify the two states where the atom is localized at one of the two ends as the eigenstates of σ^z. Experimentalists can literally take 'pictures' of the state of the system in this basis and produce a collection of pictures schematically represented in figure 10.9.

Often the energies are expressed in units of the transverse magnetic field h_T and it is common to define $\lambda \equiv J/h_t$ which has a critical value $\lambda_c = 1$. For $\lambda < 1$, we are in the disordered phase and $\lambda > 1$, we are in the ordered phase where an infinitesimal symmetry breaking can generate 'spontaneous' magnetization. In the ladder realization, h_T is proportional to the inverse tunneling time along the rungs. It is also possible to tilt the rungs and introduce a longitudinal magnetic field H. Figure 10.9 gives the three most likely measurements for $\lambda = 1.5$ and $H = 0.2$ when the system is in the ground state.

Figure 10.7. Representation of the one and two Polyakov loops (in red). Reprinted with permission of APS from [43].

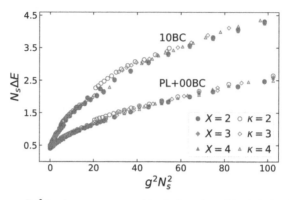

Figure 10.8. $N_s\Delta E$ versus $g^2 N_s^2$ for the gap ΔE created by the insertion of the Polaykov loop (lower set) or an external electric field (10 boundary conditions, upper set). Open (filled) markers represent Lagrangian (Hamiltonian) data. This figure is from reference [43] where the choices of parameters, units and methods for both of the 24 datasets are explained. Reprinted with permission of APS from [43].

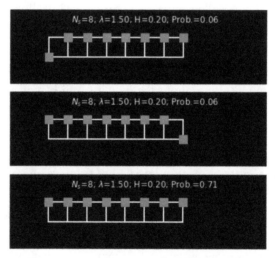

Figure 10.9. Schematic representation of σ^z measurements. The orange boxes represent the atoms. The rungs play the role of qubits.

The ladder setup that we just discussed can be imagined as an idealized quantum computer. Each rung is a qubit. We have discussed the one-qubit gates corresponding to $\hat{\sigma}^x$ and $\hat{\sigma}^z$. The nearest neighbor interactions can be used to create $\hat{\sigma}_j^x \hat{\sigma}_{j+1}^x$ two-qubit gates. The Rydberg interactions can be tuned to obtain longer range interactions. The optical lattice setup may not be practical for fast manipulations. The physical rungs may be replaced by 'synthetic' ones where the spin states correspond to atomic or molecular energy levels as, for instance, in trapped ions setups.

An interesting quantity to measure is the zero temperature magnetic susceptibility defined as

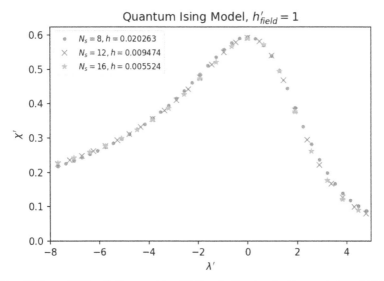

Figure 10.10. $\chi^{\text{quant.}'} = \chi^{\text{quant.}} L^{-(1-\eta)}$ versus $\lambda' = L^{1/\nu}(\lambda - 1)$. Reprinted with permission of APS from [43].

$$\chi^{\text{quant.}} = \frac{1}{L} \sum_{\langle i,j \rangle} \langle (\sigma_i - \langle \sigma_i \rangle)(\sigma_j - \langle \sigma_j \rangle) \rangle \qquad (10.27)$$

where $\langle \ldots \rangle$ are short notations for $\langle \Omega | \ldots | \Omega \rangle$ with $|\Omega\rangle$ is the ground state of $\hat{H}$. Again, it is possible to make use of results at different but not large sizes. As we will explain in section 11.5, data collapse for the quantum magnetic susceptibility can be obtained by using the rescaled quantities $\chi^{\text{quant.}'} = \chi^{\text{quant.}} L^{-(1-\eta)}$ versus $\lambda' = L^{1/\nu}(\lambda - 1)$ which are displayed in figure 10.10.

10.5 Quantum computers

Recently, commercial companies have made available actual quantum computers to the scientific community. In this subsection we discuss a specific effort designed to benchmark the performance of a few actual machines using the quantum Ising model with four sites. The details and credits can be found in reference [46]. More details will be provided in E Gustafson's dissertation.

The model is defined by the Hamiltonian (9.53) with OBC. The quantum circuit is displayed at the bottom of figure 9.7. The quantum computer provides average 'particle number' or 'qubit occupation' at each site j defined as

$$n_j = \left(1 - \langle \hat{\sigma}_j^z \rangle \right)/2. \qquad (10.28)$$

In order to compare different hardware, we defined a new metric that we have called the gross averaged discrepancy which takes all the differences of measured data points from the exact value for all data points whose measured value is above a certain threshold ϵ and averages them. This metric is algebraically defined as,

Table 10.1. Readout error probabilities from reference [46].

Qubit	Probability	Almaden	Boeblingen	Melbourne
1	$p_{1\to0}$	0.0067	0.0933	0.005
1	$p_{0\to1}$	0.0533	0.1467	0.083
2	$p_{1\to0}$	0.0200	0.0133	0.008
2	$p_{0\to1}$	0.0300	0.0900	0.028
3	$p_{1\to0}$	0.0000	0.0100	0.055
3	$p_{0\to1}$	0.0533	0.0333	0.078
4	$p_{1\to0}$	0.0066	0.0367	0.006
4	$p_{0\to1}$	0.0500	0.0267	0.068

$$G_\epsilon(\text{data set}) \equiv \frac{\sum_{n_{\text{meas}}>\epsilon} |n_{\text{meas}} - n_{\text{Trotter}}|}{\sum_{n_{\text{meas}}>\epsilon} 1}. \tag{10.29}$$

The reason for this choice is explained at length in reference [46]. This metric allows us to compare the raw data with the numerically accurate Trotter evolution, which is the best result that the quantum computer can achieve if no algorithmic mitigation is applied. It also gives a measure of the quality of readout corrections and noise mitigations.

The readout errors (misidentifying a $|1\rangle$ for a $|0\rangle$ or vice versa) for current superconducting qubit quantum computers can be as high as 15 percent. Examples for three IBM machines used in reference [46] are given in table 10.1.

Knowing the readout error probabilities, we can reconstruct the true occupations from the ones we read. Assuming that the qubit has a true probability p_0 to be in the $\sigma^z = +1$ state and probability $p_1 = 1 - p_0$ to be in the $\sigma^z = -1$ state, we have

$$\langle \hat{\sigma}^z \rangle_{\text{true}} = p_0 - p_1. \tag{10.30}$$

We can then use the readout error probabilities to obtain

$$\langle \hat{\sigma}^z \rangle_{\text{read}} = p_0(1 - p_{0\to1}) + p_1 p_{1\to0} - (p_1(1 - p_{1\to0}) + p_0 p_{0\to1}). \tag{10.31}$$

Eliminating p_0 and p_1 in terms of $\langle \hat{\sigma}^z \rangle_{\text{true}}$, yields

$$\langle \hat{\sigma}^z \rangle_{\text{read}} = \langle \hat{\sigma}^z \rangle_{\text{true}}(1 - p_{1\to0} - p_{0\to1}) + p_{1\to0} - p_{0\to1}. \tag{10.32}$$

Re-expressing in terms of $\langle \hat{\sigma}^z \rangle_{\text{read}}$, we obtain

$$\langle \hat{\sigma}^z \rangle_{\text{true}} = \frac{\langle \hat{\sigma}^z \rangle_{\text{read}} + p_{0\to1} - p_{1\to0}}{(1 - p_{0\to1} - p_{1\to0})}, \tag{10.33}$$

This scheme can be approximated by assuming that the readout errors are identical $(p_{1\to0} \simeq p_{0\to1} \simeq p_{0\leftrightarrow1})$. This yields the simplified symmetric expression:

Table 10.2. $G_0 \times 10^3$ summarized for various machines over $\delta t = 5$ from reference [46].

Machine	Raw	Symmetric	Asymmetric
Almaden	86(8)	76(9)	74(10)
Boeblingen	67(8)	51(7)	36(6)
Melbourne	120(15)	108(15)	96(11)

$$\langle \hat{\sigma}^z \rangle_{\text{true}} \simeq \frac{\langle \hat{\sigma}^z \rangle_{\text{read}}}{(1 - 2p_{0 \leftrightarrow 1})} \tag{10.34}$$

The use of the asymmetric and its symmetric approximation leads to improvements that can be measured with G_0 for the first Trotter step with $\delta t = 5$ as shown in table 10.2 which also shows that the more recent IBM machines Almaden and Boeblingen perform better than Melbourne. Using $G_{0.1}$ or $G_{0.2}$ tends to provide better discrimination [46].

Gate noise mitigation methods [47–51], inspired by Richardson [52] consist in increasing the noise in the system by fixed amounts and then extrapolating backwards to a noiseless value. This can be done by replacing a single CNOT gate by r CNOT gates to increase the noise with r odd. For a perfect quantum computer, this has no effect because CNOT is its own inverse, however, for NISQ devices, this amplifies the errors. This method gives reliable results when the amplified error is approximately linear in r [46]. Algorithmic mitigation (trying to extrapolate to zero Trotter step), is currently more difficult because, as explained in section 9.7, for $\delta t \sim 5$, the algorithmic error depends in a complicated nonlinear way on δt.

A more comprehensive list of references can be found in: Meurice Y, Sakai R and Unmuth-Yockey J Tensor field theory with applications to quantum computing arXiv:2010.06539.

References

[1] Bloch I, Dalibard J and Zwerger W 2008 *Rev. Mod. Phys.* **80** 885–964
[2] Trotzky S, Pollet L, Gerbier F, Schnorrberger U, Bloch I, Prokof'ev N V, Svistunov B and Troyer M 2010 *Nat. Phys.* **6** 998
[3] Meurice Y 2011 *PoS* **LATTICE2011** 040 https://pos.sissa.it/139/040/pdf
[4] Affleck I, Zou Z, Hsu T and Anderson P W 1988 *Phys. Rev.* B **38** 745–7
[5] Dagotto E, Fradkin E and Moreo A 1988 *Phys. Rev.* B **38** 2926–9
[6] Banerjee D, Bögli M, Dalmonte M, Rico E, Stebler P, Wiese U J and Zoller P 2013 *Phys. Rev. Lett.* **110** 125303
[7] Wiese U J 2013 *Ann. Phys.* **525** 777–96
[8] Tagliacozzo L, Celi A, Zamora A and Lewenstein M 2013 *Ann. Phys.* **330** 160–91
[9] Zohar E, Cirac J I and Reznik B 2015 *Rep. Prog. Phys.* **79** 014401
[10] Bañuls M C and Cichy K 2020 *Rep. Prog. Phys.* **83** 024401

[11] Li T, Duca L, Reitter M, Grusdt F, Demler E, Endres M, Schleier-Smith M, Bloch I and Schneider U 2016 *Science* **352** 1094–7

[12] Kasper V, Hebenstreit F, Jendrzejewski F, Oberthaler M K and Berges J 2017 *New J. Phys.* **19** 023030

[13] Fu H, Feng L, Anderson B M, Clark L W, Hu J, Andrade J W, Chin C and Levin K 2018 *Phys. Rev. Lett.* **121** 243001

[14] Schweizer C, Grusdt F, Berngruber M, Barbiero L, Demler E, Goldman N, Bloch I and Aidelsburger M 2019 *Nat. Phys.* **15** 1168–73

[15] Leibfried D, Blatt R, Monroe C and Wineland D 2003 *Rev. Mod. Phys.* **75** 281–324

[16] Davoudi Z, Hafezi M, Monroe C, Pagano G, Seif A and Shaw A 2020 2020 *Phys. Rev. Res.* **2** 023015

[17] Bernien H *et al* 2017 *Nature* **551** 579–84

[18] Keesling A *et al* 2019 *Nature* **568** 207–11

[19] Celi A, Vermersch B, Viyuela O, Pichler H, Lukin M D and Zoller P 2019 (arXiv:1907.03311)

[20] Martinez E A *et al* 2016 *Nature* **534** 516–9

[21] Kharzeev D E and Kikuchi Y 2020 *Phys. Rev. Res.* **2** 023342

[22] Klco N, Dumitrescu E F, McCaskey A J, Morris T D, Pooser R C, Sanz M, Solano E, Lougovski P and Savage M J 2018 *Phys. Rev.* A **98** 032331

[23] Surace F M, Mazza P P, Giudici G, Lerose A, Gambassi A and Dalmonte M 2020 *Phys. Rev.* X **10** 021041

[24] Magnifico G, Dalmonte M, Facchi P, Pascazio S, Pepe F V and Ercolessi E 2019 (arXiv:1909.04821)

[25] Silvi P, Sauer Y, Tschirsich F and Montangero S 2019 (arXiv:1901.04403)

[26] Raychowdhury I and Stryker J R 2020 2020 *Phys. Rev. Res.* **2** 033039

[27] Raychowdhury I and Stryker J R 2020 *Phys. Rev.* D **101** 114502

[28] Davoudi Z, Raychowdhury I and Shaw A 2020 (arXiv:2009.11802)

[29] Dasgupta R and Raychowdhury I 2020 (arXiv:2009.13969)

[30] Bañuls M C *et al* 2020 *Eur. Phys. J.* D **74** 165

[31] Islam R, Ma R, Preiss P M, Eric Tai M, Lukin A, Rispoli M and Greiner M 2015 *Nature* **528** 77–83

[32] Linke N M, Johri S, Figgatt C, Landsman K A, Matsuura A Y and Monroe C 2018 *Phys. Rev.* A **98** 052334

[33] Daley A J, Pichler H, Schachenmayer J and Zoller P 2012 *Phys. Rev. Lett.* **109** 020505

[34] Cardy J 1996 *Scaling and Renormalization in Statistical Physics, Cambridge Lecture Notes in Physics* (Cambridge: Cambridge University Press)

[35] Vidal G, Latorre J I, Rico E and Kitaev A 2003 *Phys. Rev. Lett.* **90** 227902

[36] Korepin V E 2004 *Phys. Rev. Lett.* **92** 096402

[37] Calabrese P and Cardy J L 2004 *J. Stat. Mech.* **0406** P06002

[38] Calabrese P and Cardy J L 2006 *Int. J. Quant. Inf.* **4** 429

[39] Unmuth-Yockey J, Zhang J, Preiss P M, Yang L P, Tsai S W and Meurice Y 2017 *Phys. Rev.* A **96** 023603

[40] Cardy J and Calabrese P 2010 *J. Stat. Mech.: Theory Exp.* **2010** P04023

[41] Calabrese P, Campostrini M, Essler F and Nienhuis B 2010 *Phys. Rev. Lett.* **104** 095701

[42] Bazavov A, Meurice Y, Tsai S W, Unmuth-Yockey J, Yang L P and Zhang J 2017 *Phys. Rev.* D **96** 034514

[43] Zhang J, Unmuth-Yockey J, Zeiher J, Bazavov A, Tsai S W and Meurice Y 2018 *Phys. Rev. Lett.* **121** 223201

[44] Unmuth-Yockey J, Zhang J, Bazavov A, Meurice Y and Tsai S W 2018 *Phys. Rev.* D **98** 094511

[45] Polyakov A M 1978 *Phys. Lett.* B **72** 477–80

[46] Gustafson E, Dreher P, Hang Z and Meurice Y 2019 (arXiv:1910.09478)

[47] Dumitrescu E F, McCaskey A J, Hagen G, Jansen G R, Morris T D, Papenbrock T, Pooser R C, Dean D J and Lougovski P 2018 *Phys. Rev. Lett.* **120** 210501

[48] Klco N and Savage M J 2019 *Phys. Rev.* A **99** 052335

[49] Endo S, Zhao Q, Li Y, Benjamin S and Yuan X 2018 (arXiv:1808.03623)

[50] Endo S, Benjamin S C and Li Y 2018 *Phys. Rev.* X **8** 031027

[51] Li Y and Benjamin S C 2017 *Phys. Rev.* X **7** 021050

[52] Richardson L F and Gaunt J A 1927 *Philos. Trans. R. Soc. A: Math. Phys. Eng. Sci.* **226** 299

IOP Publishing

Quantum Field Theory
A quantum computation approach
Yannick Meurice

Chapter 11

The renormalization group method

11.1 Basic ideas and historical perspective

The renormalization group (RG) approach was initially developed [1] in the context of statistical mechanics (second order phase transitions) and particle physics (asymptotic behavior, effective theories, lattice regularization). The main steps of the procedure can be summarized as follows:

  (i) Replace microscopic degrees of freedom by coarser ones (block spins, decimation, coarse graining $\cdots$).
  (ii) Rescale the new variables in such a way that the new interactions (encoded in a Hamiltonian or tensor $\cdots$) can be compared to the original ones.
 (iii) Find the fixed points of the RG transformation defined by (i) and (ii).
 (iv) Linearize the RG transformation at the fixed point; separate relevant (expanding) and irrelevant (contracting) directions. The relevant directions characterize the universal large distance behavior.

The historical importance of the RG ideas in particle physics is very significant. It has been very successful in perturbative context: non-renormalizable interactions are irrelevant. As discussed in section 1.5 this is the crucial ingredient for the development and establishment of the standard model as a predictive theory. The Brout–Englert–Higgs mechanism was considered as a viable idea because it led to a renormalizable theory of electroweak interactions. Running coupling constants connect the physics at the nuclear scale and the collider scale. Asymptotic freedom [2, 3] justifies the parton model. However, its application is more difficult at the non-perturbative level for instance in lattice field theory. The lattice is not physical in the context of high-energy physics and we need to take the continuum limit. This corresponds to fixed points of RG transformations (where the correlation lengths expressed in lattice spacing are infinite). So far numerical implementations of the RG are not used, say to extract $|V_{ub}|$ from $B \to \pi e\nu$ [4]. We rather rely on Monte Carlo simulations at different spacings and extrapolate to zero spacing. From a long term historical perspective the integral calculus analogy is that importance sampling is like quadratures (which are robust) while RG fixed points are like the fundamental theorem of calculus (initially hard but powerful when fully developed).

11.2 Coarse graining and blocking

One of the coarse graining procedures is called 'blocking' which means replacing a block of degrees of freedom by a single one. One simple method to block consists in introducing 1 in the partition function, for instance, in the following way:

$$\prod_{\mathfrak{B}} \int d\Phi_{\mathfrak{B}} \delta \left(\Phi_{\mathfrak{B}} - \sum_{x \in \mathfrak{B}} \phi_x \right) = 1, \tag{11.1}$$

where the blocks $\mathfrak{B}$ form a partition of the original lattice. For instance, on a two-dimensional square lattice, the blocks are squares with a linear size of two lattice spacing and contain four sites. The ϕ_x are the original lattice fields and $\Phi_{\mathfrak{B}}$ the block fields which receive new effective interactions after one performs the integration over the original fields. This process is illustrated in figure 11.1.

If we are able to determine an effective Hamiltonian for the block variables, we obtain RG flows in the space of Hamiltonians:

$$H \to H_{\text{eff}}^{(1)} \to H_{\text{eff}}^{(2)} \dots \tag{11.2}$$

We are interested in the fixed points of this transformation and their relevant directions. By tuning the initial Hamiltonian to be on the critical hypersurface (where all the directions are irrelevant) we can approach the fixed point. By considering initial conditions near the fixed points we can calculate the small changes under a RG transformation. This is a linear analysis. The behavior in the relevant direction has universal features (same critical exponents in apparently very different models). The RG flows are schematically represented in figure 11.2.

Computing an effective Hamiltonian by blockspinning in configuration space is difficult! Researchers who have never tried often think that it is straightforward. However, the procedure generates numerous new interactions which seem difficult to control. As an exercise, one can try to apply the blocking formula (11.1) to the two-dimensional Ising model on a square lattice. As a first step, one can start with one block with four sites in the background of eight neighbors as shown in figure 11.3. The sum of the spins in the block can take five values and the blocking can be encoded in 5×2^8 unnormalized probabilities. We can now reuse this work by assembling four such objects and summing over the spins at the center as shown in figure 11.4. We have integrated over 2^{20} configurations and obtained five block variables which can be in $5^5 = 3125$ configurations. However, the relative probabilities depend on the values of the 20 spin variables which are now at the boundary and so at this stage it does not seem possible to isolate something like nearest neighbor interactions. Pursuing this process will generate many new interactions and the task of extracting an effective Hamiltonian is very arduous.

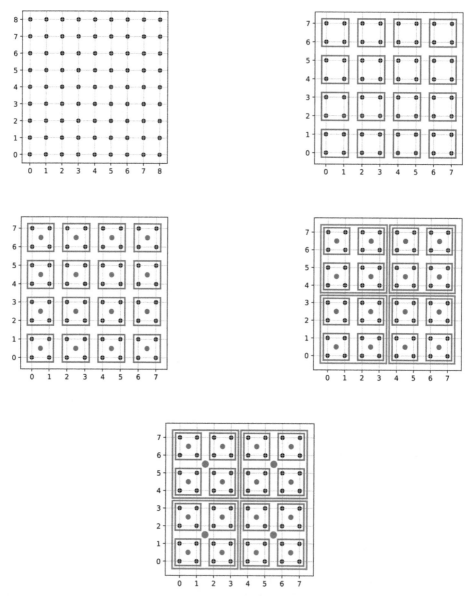

Figure 11.1. Illustration of the blocking procedure repeated twice. The blue (red) dots represent the block field variables after one (two) iterations, respectively.

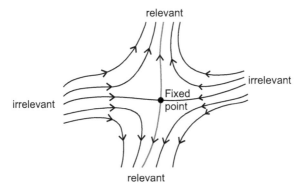

Figure 11.2. Schematic behavior of the RG flows near a fixed point with one relevant and one irrelevant direction.

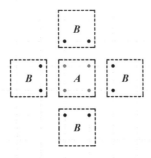

Figure 11.3. First step of a blocking procedure. Reprinted with permission of APS from [5].

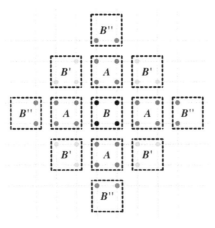

Figure 11.4. Second step of a blocking procedure. Reprinted with permission of APS from [5].

11.3 The Niemeijer–van Leeuwen equation

The complexity of the effective Hamiltonian in this approach has been acknowledged in the early RG days by Niemeijer and van Leeuwen [6]. They start with the most general Hamiltonian with couplings for products of any subset of Ising spins. If we ignore considerations such as translation invariance, there are as many couplings as configurations. This number can be reduced if we assume translation invariance. With this very general framework, we do not need to worry about generating new couplings. This framework allows formal manipulations that generate a very general equation that we will proceed to derive.

We now discuss Niemeijer and van Leeuwen (NvL) construction [6]. We follow some of their notations including the sign in front of the Hamiltonian in the partition function. For each subset of sites $\mathcal{A}$, they associate the spin

$$\sigma_{\mathcal{A}} = \prod_{x \in \mathcal{A}} \sigma_x. \tag{11.3}$$

The most general Hamiltonian is

$$\mathcal{H}(\{\sigma\}) = \sum_{\mathcal{A}} K_{\mathcal{A}} \sigma_{\mathcal{A}}. \tag{11.4}$$

The normalization condition

$$\sum_{\{\sigma\}} \mathcal{H}(\{\sigma\}) = 0, \tag{11.5}$$

is imposed. If we have V spins, we can formally invert

$$K_{\mathcal{A}} = 2^{-V} \sum_{\{\sigma\}} \sigma_{\mathcal{A}} H(\{\sigma\}). \tag{11.6}$$

From this, we see that the normalization condition only affects an additive constant in the Hamiltonian. In the following we use the notation $\mathbf{K}$ for the set of couplings. It is clear that there is a one-to-one correspondence between the domains and the spin configurations because using a domain $\mathcal{A}$, we can put spins up in $\mathcal{A}$ and spins down outside and obtain a configuration or vice versa.

Using a variation of what we just discussed in equation (11.1), NvL use block variables σ' which also take restricted values ± 1 and introduce the identity as a sum of positive probabilities

$$\sum_{\{\sigma'\}} P(\{\sigma'\}, \{\sigma\}) = 1, \tag{11.7}$$

in the partition function

$$\mathcal{Z} = \sum_{\{\sigma\}} \exp(\mathcal{H}(\{\sigma\})) = \exp(Vf(\mathbf{K})). \tag{11.8}$$

NvL define a new Hamiltonian $\mathcal{H}'(\{\sigma'\})$ using

$$\exp(G + \mathcal{H}'(\{\sigma'\})) \equiv \sum_{\{\sigma\}} P(\{\sigma'\}, \{\sigma\}) \exp(\mathcal{H}(\{\sigma\})), \tag{11.9}$$

where G depends only on the couplings and is fixed by the same normalization condition as above:

$$\sum_{\{\sigma'\}} \mathcal{H}'(\{\sigma'\}) = 0. \tag{11.10}$$

This implies that

$$G = 2^{-V'} \sum_{\{\sigma'\}} \ln \left(\sum_{\{\sigma\}} P(\{\sigma'\}, \{\sigma\}) \exp(\mathcal{H}(\{\sigma\})) \right). \tag{11.11}$$

This is a double partition function. We can find similar formal expressions for the new couplings $\mathbf{K}'$.

$$K'_A = 2^{-V'} \sum_{\{\sigma'\}} \sigma'_A \ln \left(\sum_{\{\sigma\}} P(\{\sigma'\}, \{\sigma\}) \exp(\mathcal{H}(\{\sigma\})) \right). \tag{11.12}$$

Summing over the $\{\sigma'\}$ in equation (11.9) and using equation (11.7), the partition function remains unchanged provided that we keep G in the exponent.

$$\sum_{\{\sigma'\}} \exp(\mathcal{H}'(\{\sigma'\}) + G) = \mathcal{Z}. \tag{11.13}$$

The new Hamiltonian $\mathcal{H}'(\{\sigma'\})$ has the same general parametrization as the original one. We can now define a renormalized free energy density $f'(\mathbf{K}')$ in terms of the new couplings

$$\sum_{\{\sigma'\}} \exp(\mathcal{H}'(\{\sigma'\})) = \exp(V'f'(\mathbf{K}')). \tag{11.14}$$

The new volume V' corresponds to a new lattice spacing rescaled by a factor b. For instance, $b = 2$ in figure 11.1. More generally, we will have b^D sites in a block for a D-dimensional (hyper)cubic lattice. In other words $V = V'b^D$. At this point, we need some additional assumption on $f'(\mathbf{K}')$. NvL assume that $f' = f$. From a rigorous point of view, this can't be the case because there are exponentially more couplings in V than in V'. For instance, we can consider interactions in V with more spins than V'. The fact that NvL use the same function for the free energy density in terms of the new couplings requires that V is very large and that only quasi-local couplings matter. Defining $g(\mathbf{K}) = G/V$, we obtain the NvL equation

$$f(\mathbf{K}) = g(\mathbf{K}) + b^{-D}f(\mathbf{K}'). \tag{11.15}$$

Even though computing the new couplings and the functions may be very difficult, NvL succeeded in obtaining a formal relation which can be iterated. It can also be linearized near a fixed point. It is often taken as the starting point for the introduction of the RG method in textbooks [7].

The NvL equation (11.15) provides a crucial insight on the possible singularities of the free energy. To see this let us assume that we can find a finite set of couplings, that we will keep denoting $\mathbf{K}$, are important for both the original and coarse-grained theory. In addition, let us assume that we can express the new couplings in terms of the original ones as $\mathbf{K}'(\mathbf{K})$ and that this mapping has a fixed point $\mathbf{K}^\star$. We can then study the linearized mapping $\partial \mathbf{K}'/\partial \mathbf{K}$ about the fixed point. For the sake of simplicity we assume that the eigenvalues are real, positive and that only one denoted λ is larger than one. The corresponding direction in coupling space is called a 'relevant direction' while the other directions are called 'irrelevant'. The simple case with one relevant and one irrelevant direction is illustrated in figure 11.2. Let's further assume that we can extend the relevant direction variable into a nonlinear function u of the other couplings that transforms exactly as in the linear approximation:

$$u' = \lambda u. \tag{11.16}$$

If we now ignore the irrelevant directions, the NvL equation becomes

$$f(u) = g(u) + b^{-D} f(\lambda u). \tag{11.17}$$

In addition, NvL assume that g is a regular function near the origin with an expansion

$$g(u) = \sum_{n=0}^{\infty} g_n u^n. \tag{11.18}$$

It is possible to find a regular solution of the restricted inhomogeneous NvL equation (11.17) by setting

$$f_{\text{reg.}}(u) = \sum_{n=0}^{\infty} f_n u^n. \tag{11.19}$$

For every integer such that $\lambda^n \neq b^D$, we have the formal solution

$$f_n = g_n/(1 - \lambda^n/b^D). \tag{11.20}$$

To this regular solution of the inhomogeneous equation we can add a singular solution of the homogeneous equation

$$f_{\text{sing.}}(u) = b^{-D} f_{\text{sing.}}(\lambda u). \tag{11.21}$$

A solution of this equation is

$$f_{\text{sing.}}(u) = A|u|^{\nu D}, \tag{11.22}$$

with

$$\nu = \ln b/\ln \lambda. \tag{11.23}$$

Given that $f_{\text{sing.}}(u)$ is proportional to a free energy density, the quantity $|u|^{-\nu}$ can be interpreted as a correlation length which diverges at criticality.

An interesting property of equation (11.17) is that it can be iterated.

$$f(u) = g(u) + b^{-D}[g(\lambda u) + b^{-D}f(\lambda^2 u)].$$ (11.24)

NvL assume that

$$\lim_{n \to \infty} b^{-nD} f(\lambda^n u) = 0,$$ (11.25)

and if we iterate an infinite number of times, we obtain

$$f(u) = \sum_{n=0}^{\infty} b^{-nD} g(\lambda^n u).$$ (11.26)

Exercise 1: Show that if n_0 is the first integer such that $\lambda^{n_0} > b^D$, then resummations of converging geometric series yields

$$f_{\text{sing.}}(u) = \sum_{n=-\infty}^{\infty} b^{-nD} g_{\text{rem}}(\lambda^n u),$$ (11.27)

with

$$g_{\text{rem}}(u) = \sum_{n=n_0}^{\infty} g_n u^n.$$ (11.28)

Since the series for $f_{\text{sing.}}$ now goes from minus to plus infinity, we can multiply the rescaled function by b^{-D} and shift the n index to recover the homogeneous relation.

Solution. See reference [6].

We still need to address the special situation where there is an integer $n^\star$ such that $\lambda^{n^\star} = b^D$. In this case, $u^{n^\star}$ is a regular solution of the homogeneous equation. In order to get a solution of the inhomogeneous equation, we need to introduce a logarithmic singularity. A possible solution is

$$f_{\text{sing.}} = -(g_{n^\star}/\ln(\lambda))u^{n^\star}\ln(u).$$ (11.29)

11.4 Tensor renormalization group (TRG)

Given the difficulties of blocking in configuration space, we were led to work in a much larger space of Hamiltonians where all the possible couplings were allowed and which obviously closes under blocking. The hope is that only quasi-local interactions are important as we iterate the RG transformation. However, verifying this assumption explicitly is non-trivial. In contrast, the quasi-locality is immediate in the tensor representation. It is encoded in the representation by the fact that the initial tensor has a finite number of indices, for instance $2D$ for spin models. Restricting ourselves to these tensors excludes models with long-range interactions. We will now show that this property is exactly preserved under blocking without the need to invoke the irrelevancy of highly non-local couplings.

We will explain the blocking of tensors with the simple example of the two-dimensional Ising model. We will follow some of the notations of references [8, 9] for the indices because they are geometrically suggestive, rather than the indices conventions of section 7.2. Our starting point is the partition function

$$Z = (\cosh(\beta))^{2V} \operatorname{Tr} \prod_{\text{sites}} T^{(\text{sites})}_{xx'yy'}, \tag{11.30}$$

where Tr means contractions (sums over 0 and 1) over the link indices. A remarkable feature of this representation is that the tensors can be combined easily to form another tensor. More specifically [8], this regrouping separates the degrees of freedom inside the block (integrated over), from those kept to communicate with the neighboring blocks as shown in figure 11.5. This is in contrast with the spins in configuration space that have interactions both inside and outside the block.

We now discuss the blocking of the tensors. Each of the four sites in a block has its own tensor with four indices. Two of these indices are shared with the tensors of two other sites in the block. The indices inside the block are denoted x_U, x_D, y_R, y_L in figure 11.5. If we want to follow the visual analogy with Feynman diagrams, these indices are like loop momenta. In addition we have four pair of indices which are shared with one the four neighboring blocks (which are not shown on figure 11.5). Each of these four pairs can be recombined into a single index taking four values. For instance, $X(x_2, x_2)$ is a notation for the product states, e.g. if we use Python arrays it is convenient to pick

$$X(0, 0) = 0, \ X(1, 1) = 1, \ X(1, 0) = 2, \ X(0, 1) = 3, \tag{11.31}$$

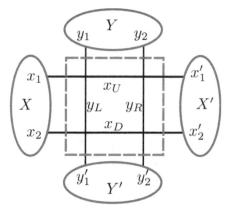

Figure 11.5. Graphical representation of the blocked tensor $T'_{XX'YY'}$ as in reference. Reprinted with permission of APS from [8].

which regroups indices with the same parity together. This blocking defines a new tensor with four indices $T'_{XX'YY'}$. These new indices have the same geometrical meaning but take twice as many values. We can now write an exact isotropic blocking formula:

$$T'_{X(x_1,x_2)X'(x_1',x_2')Y(y_1,y_2)Y'(y_1',y_2')}$$

$$= \sum_{x_U,x_D,x_R,x_L} T_{x_1 x_U y_1 y_L} T_{x_U x_1' y_2 y_R} T_{x_D x_2' y_R y_2} T_{x_2 x_D y_L y_1'},$$

The partition function can again be written as the trace of a product of tensors

$$Z \propto \mathrm{Tr} \prod_{\mathrm{sites'}} T'^{(\mathrm{sites'})}_{XX'YY'}, \tag{11.32}$$

where sites' denotes the sites of the coarser lattice with twice the lattice spacing of the original lattice. We will discuss the absolute normalization later. This process can be repeated with $T'^{(\mathrm{sites'})}_{XX'YY'}$ which has now four indices. At each step the number of indices is squared and if any kind of numerical implementation is involved, it is clear that truncations are necessary. However, it is important to realize that truncations are the *only* approximations that will be needed. The main question is what is the optimal way to do it. If we look at the way a specific tensor is connected to the rest of the system, we can define some environment tensor:

$$\mathrm{Tr} \prod_x T^{(x)} = \sum_{XX'YY'} E_{XX'YY'} T'_{XX'YY'}, \tag{11.33}$$

where $E_{XX'YY'}$ is obtained by tracing all the indices except for four pairs of indices denoted $XX'YY'$ associated with the tensor that we singled out. Calculating $E_{XX'YY'}$ is very close to what we are ultimately trying to calculate and we need to start with approximations. The main effect of the choice of the approximate form is to rank order the states of the tensor product and the explicit form of $E_{XX'YY'}$ will not appear in the approximate expressions. A very simple approximation is [8]

$$E^{\mathrm{app.}}_{XX'YY'} = C\delta_{XX'}\delta_{YY'}, \tag{11.34}$$

for some positive constant C. We can optimize the truncation by maximizing the approximate partition function expressed in terms of the trace of a matrix G such that

$$\mathrm{Tr}\, G \propto \sum_{XX'YY'} E^{\mathrm{app.}}_{XX'YY'} T'_{XX'YY'}, \tag{11.35}$$

which means we can take

$$G_{XX'} = \sum_Y T'_{XX'YY}. \tag{11.36}$$

Note that the tensor T' is symmetric under a rotation by $\pi/2$ and consequently $G_{XX'}$ is a symmetric matrix. If β is real, so is $G_{XX'}$. In addition, from the expression in terms of the original tensors, $G_{XX'}$ is in fact the square of another matrix and if β is real, then all the eigenvalues of G are positive and we can optimize the truncations by selecting the states corresponding to the largest eigenvalues of G. We will label the new indices as $0, 1, 2, \cdots$ with 0 corresponding to the largest eigenvalue, 1 for the next etc. This introduces a notion of relative importance but we can keep things exact, for instance by keeping the four indices after the first step.

We now discuss the normalization question. In the original microscopic formulation, we factored $(\cosh(\beta))^{2V}$ so that the tensor with the largest value T_{0000} was exactly one. This normalization plays the same role as the condition (11.5) in the NvL approach. Following their example, we impose the renormalization condition

$$T_{0000} = 1, \tag{11.37}$$

at each step. We have now the two steps (coarse-graining and renormalization) that define a RG transformation as discussed at the beginning of this section. Following the steps of NvL with $b = 2$ and $D = 2$ we discuss successive expressions for $\ln Z/V$. The constant term $2\ln(\cosh(\beta))$ plays no role. We can write the exact identity

$$\ln(\mathrm{Tr} \prod_{\mathrm{sites}} T^{(\mathrm{sites})}_{xx'yy'})/V = (1/4) \ln(T'_{0000}) + (1/4) \ln(\mathrm{Tr} \prod_{\mathrm{sites}'} T^{(\mathrm{sites}')}_{XX'YY'})/V'. \tag{11.38}$$

T'_{0000} is the unnormalized tensor that we just constructed. $T^{(\mathrm{sites}')}_{XX'YY'}$ is the renormalized tensor meaning the unnormalized tensor divided by T'_{0000}. Bearing in mind that $b^{-D} = 1/4$ we see some analogy with the NvL equation (11.15). $(1/4)\ln(T'_{0000})$ plays the role of $g(\mathbf{K})$. However, $\mathrm{Tr} \prod_{\cdots}$ has a different meaning in both sides of the equation. In the NvL approach a large number of couplings had to be neglected in order to write $f' = f$. Similarly, here if we decide to restrict the Tr in the rhs to a finite number of indices then the two densities have a similar form. In the approximation where we only keep 0 and 1, we obtain a mapping from a tensor with four indices taking two values into another tensor of the same type. It is not difficult to perform this mapping numerically. As we will discuss later, the argument can be extended to a larger number of states.

11.5 Critical exponents and finite-size scaling

Universal behavior is characterized by critical exponents. In the following, we provide a short discussion that should help understanding the data collapse graphs from sections 10.3 and 10.4. For a more complete discussion and a list of references see [10]. We consider the case with one relevant direction with an eigenvalue $\lambda_1 > 1$. In the linear approximation, we can use $u \sim \beta - \beta_c$ as the distance to the fixed point and the RG transformation amounts to have

$$(\beta - \beta_c) \to \lambda_1(\beta - \beta_c), \tag{11.39}$$

while the lattice spacing a increases with the scale factor b as

$$a \to ab. \tag{11.40}$$

If ξ is the correlation length in lattice spacing units, we have

$$\xi \to \frac{\xi}{b}. \tag{11.41}$$

At the critical point, the correlation length diverges as

$$\xi \propto |\beta - \beta_c|^{-\nu}. \tag{11.42}$$

This implies the relation

$$\lambda_1^\nu = b. \tag{11.43}$$

Consequently, we can estimate the critical exponent ν using an estimate of λ_1:

$$\nu = \frac{\ln(b)}{\ln(\lambda_1)}, \tag{11.44}$$

which is identical to equation (11.23).

For a D-dimensional Euclidean space–time we define the magnetic susceptibility as

$$\chi = \frac{1}{V} \sum_{<x,y>} \langle (\sigma_x - \langle \sigma_x \rangle)(\sigma_y - \langle \sigma_y \rangle) \rangle. \tag{11.45}$$

At criticality we assume the power behavior

$$\langle (\sigma_x - \langle \sigma_x \rangle)(\sigma_y - \langle \sigma_y \rangle) \rangle|_{\beta=\beta_c} \propto \frac{1}{|x - y|^{D-2+\eta}}, \tag{11.46}$$

with $\eta = 0$ for free theories as can be found by dimensional analysis. Near criticality this results holds approximately for distances smaller than ξ and becomes exponentially small for distances larger than ξ. Consequently,

$$\chi \propto \int_{r<\xi} dr r^{D-1} r^{-D+2-\eta} \propto \xi^{2-\eta}. \tag{11.47}$$

If we define the critical exponent γ for

$$\chi \propto |\beta - \beta_c|^{-\gamma}, \tag{11.48}$$

we obtain the relation

$$\gamma = \nu(2 - \eta). \tag{11.49}$$

The argument extends to the quantum case, however, we need to replace

$$\int_{r<\xi} dr r^{D-1} \to \int_{r<\xi} dr r^{D-2}, \tag{11.50}$$

and we obtain

$$\chi^{\text{quant.}} = \frac{1}{L^{D-1}} \sum_{<x,y>} \langle (\sigma_x - \langle \sigma_x \rangle)(\sigma_y - \langle \sigma_y \rangle) \rangle \propto \xi^{1-\eta}. \tag{11.51}$$

where $\langle \ldots \rangle$ are short notations for $\langle \Omega | \ldots | \Omega \rangle$ with $|\Omega\rangle$ the ground state of $\hat{H}$.

At finite volume, the linear scale L in units of lattice spacing transforms as

$$L \to L/b, \tag{11.52}$$

under a RG transformation. If a quantity Q has a behavior of the form

$$Q \propto |\beta - \beta_c|^{-A}, \tag{11.53}$$

we can construct an approximately RG-invariant quantity $QL^{-A/\nu}$. This explains the data collapse obtained by plotting $\chi^{\text{quant.}\prime} = \chi^{\text{quant.}} L^{-(1-\eta)}$ versus $\lambda' = L^{1/\nu}(\lambda - 1)$ in figure 10.10.

11.6 A simple numerical example with two states

Historically, one very new aspect brought by K Wilson's RG program is that it could be carried out numerically rather than by calculating Feynman diagrams or power series. We will show that a simple two-state approximation [8] provides reasonable estimates of the critical exponent ν. This can be explained by the fact that the the states of the transfer matrix can be classified into an even (where the sum of the indices is even) and an odd sector (where the sum of the indices is odd). In the exact solution [11], the partition function is dominated by the two leading eigenvalues (one in each sector) and the two-state approximation accommodates this feature. It also has most of the qualitative features that a more accurate treatment involving more states would have and so it is worth experimenting with the numerical implementation.

With the indices taking two values, the rank four tensor has in principle 16 independent entries, however, because of the selection rule of equation (7.15), the sum of the indices must be even and so 8 of the tensor elements are zero. In addition, we have blocked isotropically and the symmetry under the rotation of figure 11.5 by ninety degrees imposes that

$$T_{1010} = T_{0110} = T_{1001} = T_{0101} \equiv t_1, \tag{11.54}$$

and

$$T_{1100} = T_{0011} \equiv t_2. \tag{11.55}$$

In addition, we define

$$T_{1111} \equiv t_3. \tag{11.56}$$

For the initial tensor, we have

$$t_1 = t_2 = \tanh(\beta) \text{ and } t_3 = t_1^2. \tag{11.57}$$

The property $t_1 = t_2$ is not preserved by the blocking procedure which can be expressed as a mapping of the three-dimensional parameter space (t_1, t_2, t_3) into itself that we denote $t'_i(t_1, t_2, t_3)$. This was implemented with a Python code based on the simple G matrix in equation (11.36) and is provided in section 11.8. This code was written to reproduce the construction in an obvious way rather than optimizing the speed. Our next task is to find the fixed points of the RG mapping. We can get a first idea by simply plotting the values of the t_i with $i = 1, 2, 3$ for various values of β and several iterations. This is done in figure 11.6. We see that for β small enough, all the t_i become smaller and smaller as we iterate, while for β large enough, they all get close to 1. In addition, the various iterations almost cross for a value of β slightly smaller than 0.4. The transition becomes sharper and sharper as we increase the number of iterations of the RG map.

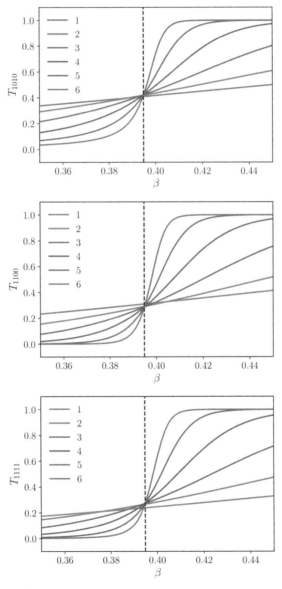

Figure 11.6. $t_1 = T_{1010}$, $t_2 = T_{1100}$ and $t_3 = T_{1111}$ versus β for six iterations of the RG mapping. The dotted line is at $\beta = 0.394\,867\,858\ldots$. As the iteration number increases, the color smoothly changes from red to blue.

By looking at the blocking formula, it is not difficult to see that $t_i = 0$ or $t_i = 1$ for $i = 1, 2, 3$ are fixed points. Empirically, we can see that their basins of attraction are separated by a critical surface. We can fine-tune β to start very close to the critical surface. If β is slightly too small, we end up on the high-temperature fixed point $t_i = 0$, if β is slightly too large, we end up on the low-temperature fixed point $t_i = 1$. If we get close enough to the critical value of β, the RG flow stays on the critical surface for a while and gets closer to a non-trivial fixed point as we iterate. We will see that this non-trivial fixed point has only one relevant direction.

As we monitored the values of a tensor for successive iterations, we noticed that the transition between the low and high temperature fixed points becomes sharper and sharper as we increase the number of iterations. We can monitor a specific tensor say T_{1100}. If after a certain number of iterations its value gets larger than say 0.8, we decide that we are in the low-temperature side. We then stop iterating and decrease β. On the other hand, if the tensor is smaller than say 0.2, we decide that we are on the high-temperature side. We then stop iterating and increase β. As we proceed, we reduce the amount by which we increase or decrease β. This can be done by multiplying $\delta\beta$ by about 0.6 at each step. This task can be performed with a Python code provided in section 11.8. If we start with an initial value $\beta = 0.39$ and a change of β of 0.005, the output is

```
0.39 (too small)
0.395 (too large)
0.392 (too small)
0.393 800 000 000 000 04 (too small)
0.394 880 000 000 000 06 (too large)
0.394 232 000 000 000 1 (too small)
0.394 620 800 000 000 1 (too small)
0.394 854 080 000 000 1 (too small)
0.394 994 048 000 000 1 (too large)
0.394 910 067 200 000 1 (too large)
0.394 859 678 720 000 1 (too small)
0.394 889 911 808 000 1 (too large)
0.394 871 771 955 200 07 (too large)
0.394 860 888 043 520 04 (too small)
........
50 more iterations
.........
0.394 867 858 135 253 (too large)
0.394 867 858 135 252 94 (too small)
0.394 867 858 135 252 94 (too small)
```

At that point, the change in β becomes smaller than the numerical precision (about 16 or 17 digits) and we cannot improve the estimate unless we increase the working precision. We conclude that the critical temperature is approximately

$$\beta_c^{2\text{states}} = 0.394\ 867\ 858\ 135\ 25....\ \ \ \ (11.58)$$

We can now look more closely at the tensor values near the crossing. This is done for t_2 in figure 11.7. As we iterate, the crossing point between two successive iterations moves towards β_c. Notice that β_c^{2states} is close but significantly different from the exact value for

$$\beta_c^{\text{exact}} = (1/2)\ln(1 + \sqrt{2}) = 0.440\,687\ldots. \tag{11.59}$$

Using our best estimate of β_c to determine the initial values from equation (11.57) we can iterate the RG mapping. As we proceed, we will get closer to the fixed point by contracting the irrelevant directions but at the same time moving away in the relevant direction. Without the knowledge of the eigenvalues of the linear transformation, it is difficult to know *a priori* the best time to stop and get an optimal estimate. Empirically we can monitor the changes in the t_i. For instance, we can calculate the norm of the difference

$$||\delta t|| \equiv \sqrt{\sum_{i=1}^{3}(t'_i - t_i)^2}. \tag{11.60}$$

This is done in a Python code available in section 11.8. The results are that $||\delta t||$ reaches a minimal value of 6×10^{-12} after 16 iterations where our estimate for the fixed point is

$$t_1^{\star} = 0.422\,298\,872\,839\,704\,85,$$
$$t_2^{\star} = 0.286\,372\,576\,051\,017\,85, \tag{11.61}$$
$$t_3^{\star} = 0.274\,663\,423\,689\,274\,4.$$

It is now possible to linearize the RG mapping near the approximate fixed point by calculating the matrix

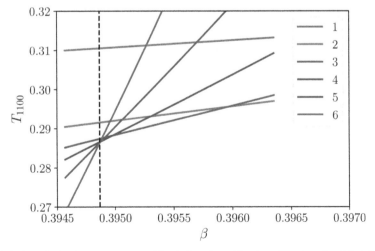

Figure 11.7. A closer look of $t_2 = T_{1100}$ versus β for six iterations. The dotted line is at $\beta = 0.394\,867\,858\ldots$. As the iteration number increases, the color smoothly changes from red to blue.

$$\mathcal{M}_{ij} = \partial t'_i/\partial t_j|_{t^\star}.$$ (11.62)

The Python code of section 11.8 shows that the eigenvalues are

$$\lambda_1 = 2.009\ 310\ 69, \quad \lambda_2 = 0.194\ 127\ 05, \quad \lambda_3 = 0.086\ 577\ 54.$$ (11.63)

As previously announced, we have one relevant direction and two irrelevant directions. Our fine-tuning of β implies that we have a closeness to the fixed point of the order of 10^{-16} in the relevant direction. After 16 iterations, this closeness becomes $\lambda_1^{16} \times 10^{-16} \sim 6 \times 10^{-12}$. At the same time if we start with differences of order 1 in the irrelevant direction, we end up with a closeness of order $\lambda_2^{16} \sim 4 \times 10^{-12}$. This is consistent with our optimal $\|\delta t\| \sim 6 \times 10^{-12}$ after 16 iterations.

The relevant scaling can be observed directly in the tensor iterations. For instance, for t_2 in figure 11.8. Near the fixed point, the distance to the fixed point in the relevant direction is essentially $\beta - \beta_c^{2\text{states}}$ and each iteration effectively multiplies this difference by $\lambda_1 \simeq 2.01$. In other words, if we start at $\beta = \beta_c^{2\text{states}} + \delta\beta$ and reach a certain value of t_2 after ℓ iterations, we would reach approximately the same value after $\ell + 1$ iterations by starting at $\beta = \beta_c^{2\text{states}} + \delta\beta/\lambda_1$. This mechanism is illustrated in figure 11.8 with $\delta_{\text{beta}} = 0.001\ 1$.

More generally, it is possible to obtain a complete data collapse of the previous figures by rescaling $\beta - \beta_c^{2\text{states}}$ by $\lambda_1^{-\ell}$ after ℓ iterations. This is illustrated in figure 11.9.

Despite its simplicity, the two-state approximation provides good estimates of the critical exponent ν. Using the definition (11.23) with our estimate of λ_1 we obtain

$$\nu \simeq 0.993,$$ (11.64)

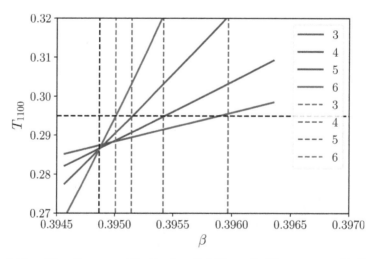

Figure 11.8. Scaling of $t_2 = T_{1100}$ versus β for iterations 3–6. The vertical lines correspond to $\delta_{\text{beta}} = 0.001\ 1$, $0.000\ 55$, $0.000\ 275$, and $0.000\ 14$, from right to left.

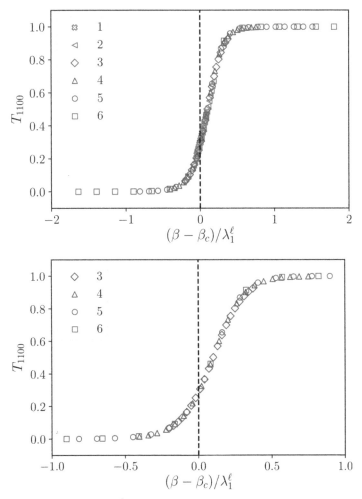

Figure 11.9. $t_2 = T_{1100}$ versus $(\beta - \beta_c)/\lambda_1^\ell$ for $\ell = 1, \dots 6$ iterations of the two-state approximation. Iterations $\ell = 3\dots6$ are shown with a magnified scale for the bottom graph.

which is surprisingly close to the exact value $\nu = 1$ and much better than the Migdal–Kadanoff approximations [12, 13] which both provide $\nu = 1.338$. Similar numerical results are obtained with a dual construction in reference [14].

A more refined approximation inspired by reference [15] discussed in section 11.7 would be to take the environment as a 'mirror image' of the tensor itself

$$E^{\text{app.}}_{XX'YY} = C' T'^{(0)\star}_{XX'YY'}. \tag{11.65}$$

This method allows us to handle the case of complex temperatures. The trace of G can then be identified with the tensor norm:

$$\text{Tr}\, G = \sum_{XX'YY'} T'_{XX'YY'} T'^{\star}_{XX'YY'} = \|T'\|^2. \tag{11.66}$$

This is clearly a sum of positive terms Truncation can be accomplished by selecting the eigenstates of Hermitian matrix

$$G_{XX'} = \sum_{X''YY'} T'_{XX''YY'} T'^{\star}_{X'X''YY'}, \tag{11.67}$$

with the largest eigenvalues.

11.7 Numerical implementations

We can attempt to improve the two-state approximation that we just discussed by allowing more states. For the Ising model, we could proceed exactly, without truncations, until we reach say 16 states and then reduce the tensor product of dimension 256 to a smaller number of states d_s. In practical implementations, memory and CPU time considerations impose restrictions on d_s. If we look at figure 11.5, we will need to store d_s^8 values of the blocked tensor in memory before truncating. Each of these calculations involves four sums over the internal indices and as a first estimate, the CPU cost of each calculation scales like d_s^4. However, because of the symmetry of the blocked tensor we divide the graph into two identical parts, say one on the left and one on the right. These two identical parts have the structure shown in figure 11.10 and the memory scales like d_s^6 while the summation over the internal index scales like d_s. The total CPU time for this part of the calculation scales like d_s^7. We can now assemble two of these tensors to build the matrix G in equation (11.36) which has d_s^4 entries, each of them requiring four internal sums and the CPU naively scales like d_s^8. In this case, G is itself a square and the cost can be reduced to d_s^5.

It is more efficient to use an anisotropic method where one proceeds as for the subgraphs discussed above and then project the two tensor products of dimension d_s^2 into a d_s subspace. This method was originally proposed in reference [15] where computations with $d_s = 24$ were performed for the two-dimensional Ising model with an accuracy of at least six significant digits for the free energy. Their anisotropic blocking involving two sites i and j provides a new rank-4 tensor:

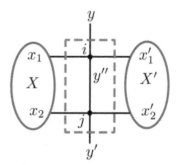

Figure 11.10. Graphical representation of $M^{<ij>}_{X(x_1, x_2)X'(x_1'x_2')yy'}$, the boundary of the block is represented by the dashed lines and the product states by ovals, as in reference [8]. Reprinted with permission of APS from [8].

$$M^{<ij>}_{X(x_1, x_2)X'(x_1'x_2')yy'} = \sum_{y''} T^{(i)}_{x_1, x_1', y, y''} T^{(j)}_{x_2x_2'y''y'}, \qquad (11.68)$$

which is represented graphically in figure 11.10. The truncation can be performed by maximizing the norm of the tensor M which can be expressed as the trace of the matrix

$$G_{XX'} = \sum_{X''yy'} M_{XX''yy'} M^*_{X'X''yy'}, \qquad (11.69)$$

because

$$\mathrm{Tr}\, G = \sum_{XX'yy'} M_{XX'yy'} M^*_{XX'yy'} = ||M||^2. \qquad (11.70)$$

The cost of computing each entry of G is d_s^4 coming from $\sum_{X''yy'}$ and the CPU scales like d_s^8. In summary, for numerical purposes in two dimensions, it is desirable to split the isotropic blocking into two steps. However, if we are looking for fixed points of the isotropic RG, we need to consider two steps.

At finite volume, a better description of the environment can be reached by following a local truncation procedure until there is no environment and equation (11.34) can be used as exact. It is then possible to move 'backward' [16, 17] and reconstitute the approximate environment of a single tensor coarse-grained one less time [9]. The analogy with the backward propagation used in machine learning has been exploited to design new algorithms recently [18]. More generally, algorithmic improvement in the TRG context is a subject of active investigation [19–21].

11.8 Python code

This is a pdf of a shortened version of the Python notebook used at various places in this chapter. It implements the procedure discussed in the text in a straightforward way and is not optimized for speed.

trgshort

June 28, 2020

```
[1]:  #importing python libraries
      import scipy.special
      import sys
      %matplotlib inline
      import numpy as np
      import matplotlib as mpl
      import matplotlib.pyplot as plt
      import os
      import datetime
      import math
      import itertools
      from numpy import linalg as LA
      from matplotlib import rc


      #global definitions of indices
      #relabeling the two-dimensional tensor product as a one dimensional array
      #the first two new indices have an even parity, the last two an odd parity
      ii=np.zeros((2,2),dtype=np.int)
      ii[0,0]=0
      ii[1,1]=1
      ii[1,0]=2
      ii[0,1]=3


      #this is the main iteration: it takes the three independent tensor values
      #and returns the three new ones
      def it2st(ttt1,ttt2,ttt3):

          TT=np.zeros((2,2,2,2))
          TV=np.zeros((4,4,2,2))
          TT4=np.zeros((4,4,4,4))
          GG=np.zeros((4,4))
          newTT=np.zeros((2,2,2,2))
          global ii
```

```
#initialize the tensor assuming 90 degree symmetry
#use prb87 convention
TT[0,0,0,0]=1.0
TT[1,0,1,0]=TT[1,0,0,1]=TT[0,1,1,0]=TT[0,1,0,1]=ttt1
TT[0,0,1,1]=TT[1,1,0,0]=ttt2
TT[1,1,1,1]=ttt3


#one contraction
#some line breaks are introduced to fit the width of the page

for xx1,xx1p,xx2,xx2p,yy,yyp in itertools.product(range(2),
                           range(2),range(2),range(2),range(2),range(2)):
    for ysum in range(2):
        tempo=(TT[xx1, xx1p, yy, ysum]*TT[xx2, xx2p, ysum, yyp])
        TV[ii[xx1, xx2], ii[xx1p, xx2p], yy, yyp]+=tempo
        pass
    pass

#assembling the isotropic four leg tensor

for ii1,ii2,yy,zz,yyp,zzp in itertools.product(range(4),
                   range(4),range(2),range(2),range(2),range(2)):
    for iisum in range(4):
        tempo=(TV[ii1, iisum, yy, yyp]*TV[iisum, ii2, zz, zzp])
        TT4[ii1,ii2,ii[yy, zz],ii[yyp, zzp]]+=tempo
        pass
    pass
#tracing to get the matrix G

for ii1,ii2 in itertools.product(range(4),range(4)):
    for iisum in range(4):
        GG[ii1,ii2]+=TT4[ii1,ii2,iisum,iisum]

#eigenvectors of symmetric (hermitian) matrix G
ei=LA.eigh(GG)

#new tensor, project on the two eigenvectors corresponding to the
#largest eigenvalues of G
#ei[0]: eigenvalues in INCREASING order
#ei[1]: eigenvectors[vector indices,eigenvalue label]

for jj,kk,ll,mm in itertools.product(range(2),range(2),range(2),range(2)):
    for ii1,ii2,ii3,ii4 in itertools.product(range(4),
                        range(4),range(4),range(4)):

        tempo=(TT4[ii1,ii2,ii3,ii4]*ei[1][ii1,3-jj]*
```

```
                              ei[1][ii2,3-kk]*ei[1][ii3,3-ll]*ei[1][ii4,3-mm])
                    newTT[jj,kk,ll,mm]+=tempo
                    pass
                pass
            #renormalization by tensor with 4 indices
            #corresponding to the largest eigenvalue
            TT=np.array(newTT/newTT[0,0,0,0])
            #new tensor values with prb87 conventions
            tt1p=TT[1,0,1,0]
            tt2p=TT[1,1,0,0]
            tt3p=TT[1,1,1,1]
            return([tt1p,tt2p,tt3p])
```

[2]:
```
beta=0.39 #guess for critical beta
delb=0.005 #change in beta
for finet in range(70):

    #iteration loop
    indi=0.5 #used to monitor the side
    nbl=0 #number of blocking  iterations
    #initial values of the tensors
    tt=np.tanh(beta)
    tt1=tt
    tt2=tt
    tt3=tt*tt
    while indi<0.8 and indi>0.2:
        newtt=it2st(tt1,tt2,tt3) #RG map followed by update
        indi=newtt[0]
        tt1=newtt[0]
        tt2=newtt[1]
        tt3=newtt[2]
        nbl+=1
        #to interupt in case of slow convergence
        if nbl>=100:
            print("ouch")
            break
        pass
    if indi>0.8:
        print(beta,"(too large)")
        beta-=delb #decrease beta
    else:
        print(beta,"(too small)")
        beta+=delb #increase beta
    delb*=0.6 #reduce the change magnitude

    pass
print("estimated beta critical:",beta)

betac=beta
```

```
0.39 (too small)
0.395 (too large)
0.392 (too small)
0.39380000000000004 (too small)
0.39488000000000006 (too large)
0.3942320000000001 (too small)
0.3946208000000001 (too small)
0.3948540800000001 (too small)
0.3949940480000001 (too large)
0.3949100672000001 (too large)
0.3948596787200001 (too small)
0.3948899118080001 (too large)
0.39487177195520007 (too large)
0.39486088804352004 (too small)
0.394867418390528 (too small)
0.3948713365987328 (too large)
0.3948689856738099 (too large)
0.3948675751188562 (too small)
0.39486842145182843 (too large)
0.3948679136520451 (too large)
0.3948676089721751 (too small)
0.3948677917800971 (too small)
0.39486790146485035 (too large)
0.3948678356539984 (too small)
0.3948678751405096 (too large)
0.39486785144860287 (too small)
0.3948678656637469 (too large)
0.3948678571346605 (too small)
0.39486786225211234 (too large)
0.39486785918164125 (too large)
0.3948678573393586 (too small)
0.39486785844472816 (too large)
0.3948678577815064 (too small)
0.3948678581794395 (too large)
0.39486785794067963 (too small)
0.39486785808393554 (too small)
0.39486785816988906 (too large)
0.39486785811831693 (too small)
0.3948678581492602 (too large)
0.3948678581306942 (too small)
0.3948678581418338 (too large)
0.39486785813515 (too small)
0.39486785813916025 (too large)
0.3948678581367541 (too large)
0.39486785813531045 (too large)
```

```
0.39486785813444425 (too small)
0.394867858134964 (too small)
0.39486785813527586 (too large)
0.39486785813508873 (too small)
0.394867858135201 (too small)
0.3948678581352683 (too large)
0.3948678581352279 (too small)
0.39486785813525216 (too small)
0.3948678581352667 (too large)
0.394867858135258 (too large)
0.39486785813525277 (too small)
0.39486785813525593 (too large)
0.39486785813525405 (too large)
0.39486785813525294 (too small)
0.3948678581352536 (too large)
0.3948678581352532 (too large)
0.39486785813525253 (too large)
0.3948678581352528 (too small)
0.39486785813525294 (too small)
0.39486785813525253 (too large)
0.39486785813525294 (too small)
0.39486785813525294 (too small)
0.39486785813525294 (too small)
0.39486785813525294 (too small)
0.39486785813525294 (too small)
estimated beta critical: 0.39486785813525294
```

[3]:
```
#change in beta at the end
delb
```

[3]: 1.4776022072738366e-18

[4]:
```
#initial values with beta critical
tt=np.tanh(betac)
nbl=0
tt1=tt
tt2=tt
tt3=tt*tt
nbl=1
old=[tt1,tt2,tt3]
fpestimate=[tt1,tt2,tt3]
change=1.0
bestiterate=1
#RG iterations
while nbl<=20:
    if nbl>=10:
        print("iteration:",nbl,"initial tensor values:",tt1,tt2,tt3)
```

```
    newtt=it2st(tt1,tt2,tt3)
    tt1=newtt[0]
    tt2=newtt[1]
    tt3=newtt[2]
    diff=np.array(newtt)-np.array(old) #changes in values of t_i
    if nbl>=10:
        print("tensor differences:",diff)
    difnorm=np.linalg.norm(diff)
    #print(" difference norm",difnorm)
    if (difnorm<change):
        bestiterate=nbl
        fpestimate=newtt
        change=difnorm
    old=newtt
    nbl+=1
    pass
print("the tensor is closest to the fixed point after",bestiterate,"iterations")
print("estimated fixed point:",fpestimate)
ttstar=fpestimate
```

iteration: 10 initial tensor values: 0.42229884428267334 0.2863726311179056
0.2746633439115711
tensor differences: [2.30175742e-08 -4.43727954e-08 6.42954874e-08]
iteration: 11 initial tensor values: 0.42229886730024757 0.2863725867451102
0.27466340820705853
tensor differences: [4.46825693e-09 -8.61402683e-09 1.24814020e-08]
iteration: 12 initial tensor values: 0.4222988717685045 0.28637257813108336
0.2746634206884605
tensor differences: [8.67267980e-10 -1.67237429e-09 2.42282922e-09]
iteration: 13 initial tensor values: 0.4222988726357725 0.28637257645870906
0.2746634231112897
tensor differences: [1.68078385e-10 -3.24977156e-10 4.70046169e-10]
iteration: 14 initial tensor values: 0.42229887280385087 0.2863725761337319
0.2746634235813359
tensor differences: [3.20627969e-11 -6.37390141e-11 9.06666964e-11]
iteration: 15 initial tensor values: 0.42229887283591366 0.2863725760699929
0.2746634236720026
tensor differences: [5.08754150e-12 -1.36843870e-11 1.64315228e-11]
iteration: 16 initial tensor values: 0.4222988728410012 0.2863725760563085
0.2746634236884341
tensor differences: [-1.29635191e-12 -5.29065680e-12 8.40272296e-13]
iteration: 17 initial tensor values: 0.42229887283970485 0.28637257605101785
0.2746634236892744
tensor differences: [-4.84107199e-12 -6.31961150e-12 -4.55774307e-12]
iteration: 18 initial tensor values: 0.4222988728348638 0.28637257604469823
0.27466342368471663
tensor differences: [-1.01607611e-11 -1.18612342e-11 -1.03705378e-11]

iteration: 19 initial tensor values: 0.422298872824703 0.286372576032837
0.2746634236743461
tensor differences: [-2.05012118e-11 -2.36705100e-11 -2.10730877e-11]
iteration: 20 initial tensor values: 0.4222988728042018 0.2863725760091665
0.274663423653273
tensor differences: [-4.12095913e-11 -4.75304240e-11 -4.23883151e-11]
the tensor is closest to the fixed point after 16 iterations
estimated fixed point: [0.42229887283970485, 0.28637257605101785,
0.2746634236892744]

[6]:
```
#best estimate of fixed point
ttstar
```

[6]: [0.42229887283970485, 0.28637257605101785, 0.2746634236892744]

[7]:
```
#checking the fixed point
np.array(it2st(ttstar[0],ttstar[1],ttstar[2]))-np.array(ttstar)
```

[7]: array([-4.84107199e-12, -6.31961150e-12, -4.55774307e-12])

[8]:
```
#jac: linearized matrix at the estimated fixed point
tt1star=ttstar[0]
tt2star=ttstar[1]
tt3star=ttstar[2]
#this is a simple linear implementation of the derivative
deltaa=10**(-6)
jac1=(np.array(it2st(tt1star+deltaa,tt2star,tt3star))-np.array(ttstar))/deltaa
jac2=(np.array(it2st(tt1star,tt2star+deltaa,tt3star))-np.array(ttstar))/deltaa
jac3=(np.array(it2st(tt1star,tt2star,tt3star+deltaa))-np.array(ttstar))/deltaa
jac=[jac1,jac2,jac3]
```

[9]:
```
jac
```

[9]: [array([1.14588301, 1.14911128, 1.13122757]),
       array([0.6515363 , 0.92804922, 0.61862377]),
       array([0.10892019, 0.09512472, 0.21608304])]

[10]:
```
eigenv=LA.eig(jac)
```

[11]:
```
eigenv
```

[11]: (array([2.00931069, 0.08657754, 0.19412705]),
       array([[-0.83300259, -0.82607279, 0.60641181],
              [-0.54750725, 0.28009231, -0.75318931],
              [-0.07963979, 0.48903175, 0.25489325]]))

[12]:
```
#nu
np.log(2)/np.log(eigenv[0][0])
```

[12]: 0.993343943233837

[13]:
```
#omega
-np.log(eigenv[0][2])/np.log(2)
```

[13]: 2.3649269444571006

[14]:
```
nte=100 #number of temperatures
nit=6 #number of iterations
#these array will store a tensor at each itertion for the set of betas
tfg1=np.zeros((nte,nit))
tfg2=np.zeros((nte,nit))
tfg3=np.zeros((nte,nit))
tlis=np.zeros(nte) #this array will store the betas
#ttin=np.zeros(nte)


#beta loop
for jjj in range(nte):
    beta=betac-0.05+0.0015*jjj
    tlis[jjj]=beta
    tt=np.tanh(beta)
    tt1=tt
    tt2=tt
    tt3=tt*tt
    for niter in range(nit):
        newtt=it2st(tt1,tt2,tt3)
        tt1=newtt[0]
        tt2=newtt[1]
        tt3=newtt[2]
        tfg1[jjj,niter]=tt1
        tfg2[jjj,niter]=tt2
        tfg3[jjj,niter]=tt3

        pass
```

[15]:
```
#plotting
mylocations=('upper left','upper left','upper left','upper center')
mylinestyle=('solid','dashed','dotted')
SMALL_SIZE = 14
MEDIUM_SIZE = 16
BIGGER_SIZE = 18

plt.rc('font', size=MEDIUM_SIZE)        # controls default text sizes
```

```
plt.rc('axes', titlesize=BIGGER_SIZE)      # fontsize of the axes title
plt.rc('axes', labelsize=MEDIUM_SIZE)      # fontsize of the x and y labels
plt.rc('xtick', labelsize=SMALL_SIZE)      # fontsize of the tick labels
plt.rc('ytick', labelsize=SMALL_SIZE)      # fontsize of the tick labels
plt.rc('legend', fontsize=SMALL_SIZE)      # legend fontsize
plt.rc('figure', titlesize=BIGGER_SIZE)    # fontsize of the figure title
plt.rc('axes', linewidth=1.)
fig = plt.figure()

plt.rc('text', usetex=True)
plt.rc('font', family='Times')
plt.ylabel('$T_{1010}$')
plt.xlabel(r'$\beta$')
colors=("b","y","g","m","r")
markers=("s","o","D","^","o","+")

for hhh in range(nit):
    plt.plot(tlis,tfg1[:,hhh],color=(1-0.15*hhh,0.1,0.15*hhh),linewidth=2)
    pass
plt.axis([0.35, 0.45, -0.1,1.1])
plt.legend(('1','2', '3', '4','5','6'),
          loc=mylocations[1])
plt.axvline(betac, color=(0.0,0.0,0.0), linestyle='--')
plt.savefig("t1010.pdf",bbox_inches='tight')
plt.show()
```

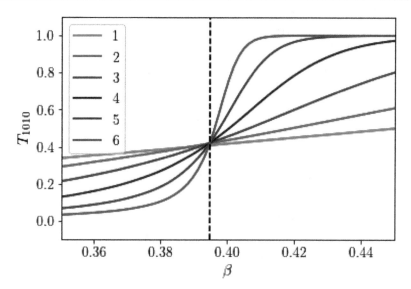

11.9 Additional material

In the following we briefly discuss more advanced material and provide references that can be consulted for more information.

Numerical methods based on singular value decomposition (SVD) were proposed in the early TRG references [22–24]. By using the SVD, the tensor T that is regarded as a $d_s^2 \times d_s^2$ matrix corresponding to the tensor product of opposite corners and can be exactly decomposed with d_s^2 principal values and approximated by keeping the d_s largest ones:

$$T_{xx'yy'} = M_{(xy)(x'y')} = \sum_{m=1}^{d_s^2} S^{[1]}_{(xy)m} \lambda_m^{[12]} S^{[2]}_{m(x'y')} \simeq \sum_{m=1}^{d_s} S^{[1]}_{(xy)m} \lambda_m^{[12]} S^{[2]}_{m(x'y')}, \qquad (11.71)$$

where $\lambda^{[12]}$ are the eigenvalues of the SVD. With this method, the reduction of the Hilbert space dimension takes place before the integration over the short distance degrees of freedom which is performed by assembling together four corners into a small scale square with d_s indices in each of the four internal leg and the four external legs. The calculation can be decomposed into two steps by first cutting the square in two identical parts, summing over one internal leg with a CPU cost of order d_s^5 and then assembling the two halves at a CPU cost of order d_s^6. The new tensor is rotated by 45 degrees and the lattice spacing increases by a factor $\sqrt{2}$. When repeated, the lattice has the original orientation and a lattice spacing twice larger.

Concerning the improvement of the two-state approximation, when a few more states are added, we noticed that the quality of the approximation does not immediately improve. One first observes oscillations, false bifurcations, approximate degeneracies. Similar observations were made by other researchers and were documented in reference [25] (see figure 2.5 therein). It has been realized in early work that the TRG has non-physical fixed points related to the so-called corner double line (CDL) tensor [26]. Taking this into account, improved coarse-graining algorithms have been proposed [27–30]. A detailed understanding about how the algorithm handles the entanglement at different scales is crucial.

The blocking procedure for gauge theories with three space–time dimensions is discussed in reference [5]. The A and B tensors do not close under blocking: if we try to combine the B tensors of two adjacent plaquettes in the same plane into a new one, this does not work because the A tensor on the common link induces two new legs orthogonal to the plane and pointing in opposite directions. We can modify the B tensor to form a $\tilde{B}$ tensor with six indices and initial value

$$\tilde{B}_{n_1 n_2 n_3 n_4 zz'} = B_{n_1 n_2 n_3 n_4} \delta_{zz'},$$

for a plaquette in the x–y plane and with similar expressions for the two other planes. The new legs piercing the plaquettes can be traced by introducing a new tensor $C_{xx'yy'zz'}$ at the center of the cubes with initial value

$$C_{xx'yy'zz'} = \delta_{xx'} \delta_{yy'} \delta_{zz'},$$

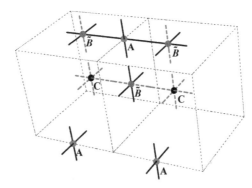

Figure 11.11. Blocking procedure. Reprinted with permission of APS from [5].

We can now rewrite the partition function as

$$Z = K(2\cosh\beta)^{3V}\,\mathrm{Tr}\,\prod_l A^{(l)} \prod_p \tilde{B}^{(p)} \prod_c C^{(c)},$$

A blocking procedure can be constructed by sequentially combining two cubes into one in each of the directions as illustrated in figure 11.11. Explicit blocking formulas can be found in reference [5].

Recently, progress has been made to reduce the algorithmic complexity in three and four space–time dimensions. The anisotropic tensor renormalization group (ATRG) [31] reduces the CPU time and the memory resource to a scaling $\mathcal{O}\!\left(d_s^{2D+1}\right)$ and $\mathcal{O}\!\left(d_s^{D+1}\right)$, respectively. See [32–34]. Another approach is to coarse-grain on a triad tensor network representation [35].

References

[1] Wilson K G and Kogut J B 1974 *Phys. Rep.* **12** 75–199
[2] Gross D J and Wilczek F 1973 *Phys. Rev. Lett.* **30** 1343–6
[3] Politzer H D 1973 *Phys. Rev. Lett.* **30** 1346–9
[4] Bailey J A *et al* 2015 *Phys. Rev. D* **92** 014024
[5] Liu Y, Meurice Y, Qin M P, Unmuth-Yockey J, Xiang T, Xie Z Y, Yu J F and Zou H 2013 *Phys. Rev. D* **88** 056005
[6] Niemeijer T and van Leeuwen J 1976 Renormalization: Ising-like spin systems *Phase Transitions and Critical Phenomena* ed C Domb and M Green vol 6 (New York: Academic)
[7] Cardy J 1996 *Scaling and Renormalization in Statistical Physics* (Cambridge Lecture Notes in Physics) (Cambridge: Cambridge University Press)
[8] Meurice Y 2013 *Phys. Rev. B* **87** 064422
[9] Xie Z Y, Chen J, Qin M P, Zhu J W, Yang L P and Xiang T 2012 *Phys. Rev. B* **86** 045139
[10] Meurice Y 2007 *J. Phys.* **A40** R39
[11] Kaufman B 1949 *Phys. Rev.* **76** 1232–43
[12] Migdal A A 1975 *Sov. Phys. JETP* **42** 743 http://www.jetp.ac.ru/cgi-bin/dn/e_042_04_0743.pdf
Migdal A A 1975 *Zh. Eksp. Teor. Fiz.* **69** 1457

[13] Kadanoff L P 1976 *Ann. Phys.* **100** 359–94

[14] Aoki K I, Kobayashi T and Tomita H 2009 *Int. J. Mod. Phys.* B **23** 3739–51

[15] Xie Z Y, Chen J, Qin M P, Zhu J W, Yang L P and Xiang T 2012 *Phys. Rev.* B **86** 045139

[16] Xie Z Y, Jiang H C, Chen Q N, Weng Z Y and Xiang T 2009 *Phys. Rev. Lett.* **103** 160601

[17] Zhao H H, Xie Z Y, Chen Q N, Wei Z C, Cai J W and Xiang T 2010 *Phys. Rev.* B **81** 174411

[18] Chen B B, Gao Y, Guo Y B, Liu Y, Zhao H H, Liao H J, Wang L, Xiang T, Li W and Xie Z Y 2020 *Phys. Rev.* B **101** 220409

[19] Bal M, Mariën M, Haegeman J and Verstraete F 2017 *Phys. Rev. Lett.* **118** 250602

[20] Fishman M T, Vanderstraeten L, Zauner-Stauber V, Haegeman J and Verstraete F 2018 *Phys. Rev.* B **98** 235148

[21] Morita S and Kawashima N 2020 (arXiv: 2009:01997)

[22] Levin M and Nave C P 2007 *Phys. Rev. Lett.* **99** 120601

[23] Jiang H C, Weng Z Y and Xiang T 2008 *Phys. Rev. Lett.* **101** 090603

[24] Gu Z C, Levin M and Wen X G 2008 *Phys. Rev.* B **78** 205116

[25] Efrati E, Wang Z, Kolan A and Kadanoff L P 2014 *Rev. Mod. Phys.* **86** 647–67

[26] Gu Z C and Wen X G 2009 *Phys. Rev.* B **80** 155131

[27] Evenbly G and Vidal G 2015 *Phys. Rev. Lett.* **115** 180405

[28] Yang S, Gu Z C and Wen X G 2017 *Phys. Rev. Lett.* **118** 110504

[29] Hauru M, Delcamp C and Mizera S 2018 *Phys. Rev.* B **97** 045111

[30] Evenbly G 2018 *Phys. Rev.* B **98** 085155

[31] Adachi D, Okubo T and Todo S 2019 (arXiv: 1906.02007)

[32] Akiyama S, Kuramashi Y, Yamashita T and Yoshimura Y 2019 Phase transition of four-dimensional Ising model with tensor network scheme *37th Int. Symp. on Lattice Field Theory (Lattice 2019) (Wuhan, Hubei, China, June 16–22, 2019)*

[33] Akiyama S, Kadoh D, Kuramashi Y, Yamashita T and Yoshimura Y 2020 (arXiv: 2005.04645)

[34] Akiyama S, Kuramashi Y, Yamashita T and Yoshimura Y 2020 (arXiv: 2009.11583)

[35] Kadoh D and Nakayama K 2019 (arXiv: 1912.02414)

IOP Publishing

Quantum Field Theory
A quantum computation approach
Yannick Meurice

Chapter 12

Advanced topics

12.1 Lattice equations of motion

In this subsection, we discuss the connection between the lattice equations of motion and the tensor selection rules. We follow the presentation of reference [1]. In section 2.1, we explained that in classical field theory, the classical equations of motion have to be used in order to establish Noether's theorem, in other words, the equality between two boundary terms: the initial and final Noether charge. In chapter 8, we showed that the selection rules appearing in the tensor elements were the consequences of symmetries. In contrast to the treatment of symmetries in conventional formulation of field theory in configuration space, there is no sharp distinction between discrete and continuous symmetries and the statement also applies to discrete symmetries. These selection rules can be seen as orthogonality relations which appear when we integrate over the fields after the character expansion and the same feature appears if the integrations are replaced by sums. In contrast, the equations of motion require continuous variations.

We will now discuss these questions for the compact abelian Higgs model (CAHM) considered in section 7.6. The lattice action is

$$S_{\text{CAHM}} = \beta_{\text{pl.}} \sum_{x,\mu<\nu} (1 - \cos(A_{x,\mu} + A_{x+\hat{\mu},\nu} - A_{x+\hat{\nu},\mu} - A_{x,\nu})) \tag{12.1}$$

$$+\beta_{l.} \sum_{x,\mu} (1 - \cos(\varphi_{x+\hat{\mu}} - \varphi_x + A_{x,\mu})). \tag{12.2}$$

We can obtain the pure gauge limit by setting $\beta_{l.} = 0$ and the $O(2)$ limit by setting the $A_{x,\mu}$ variables to zero. The variations with respect to φ_x provide the first set of equations of motion

$$\frac{\partial S_{\text{CAHM}}}{\partial \varphi_x} = \beta_l \sum_\mu [-\sin(d_{x,\mu}) + \sin(d_{x-\hat{\mu},\mu})] = 0, \tag{12.3}$$

with the notation

$$d_{x,\mu} \equiv \varphi_{x+\hat{\mu}} - \varphi_x + A_{x,\mu}, \tag{12.4}$$

which approximates the covariant derivative of φ.

We have shown in section 7.3 that the integration with respect to φ_x implies

$$\sum_\mu [-n_{x,\mu} + n_{x-\hat{\mu},\mu}] = 0. \tag{12.5}$$

The two above equations have identical geometrical structures and can be obtained from each other by the substitution

$$\beta_l \sin(d_{x,\mu}) \leftrightarrow n_{x,\mu}. \tag{12.6}$$

In equation (12.5), the $n_{x,\mu}$ come with a minus (links coming 'out' of x in the positive directions), while the $n_{x-\hat{\mu},\mu}$ come with a plus (links coming 'in' the site x from the negative direction). This feature is dictated by the sign convention appearing in the Fourier expansion. As signs are important, let's discuss this more precisely.

The conventions that we used in chapter 7 are that the same signs that appear for the linear combinations of angles inside the cosines and those multiplying the Fourier modes.

$$e^{\beta_l \cos(\varphi_{x+\hat{\mu}} - \varphi_x + A_{x,\mu})} = \sum_{n_{x,\mu}=-\infty}^{+\infty} e^{i n_{x,\mu}(\varphi_{x+\hat{\mu}} - \varphi_x + A_{x,\mu})} I_{n_{x,\mu}}(\beta_l), \tag{12.7}$$

When we perform the integration over φ_x, this variable appears in $2D$ links. It appears D times as

$$e^{i n_{x,\mu}(\cdots - \varphi_x \cdots)}, \tag{12.8}$$

and D times as

$$e^{i n_{x-\hat{\mu},\mu}(+\varphi_{x-\hat{\mu}+\hat{\mu}} \cdots)}. \tag{12.9}$$

From this we recognize the signs of equation (12.5). Similarly, when we take the derivative with respect to φ_x, we have for the same links

$$\frac{\partial(\cdots - \beta_l \cos(\cdots - \varphi_x \cdots))}{\partial \varphi_x} = -\beta_l \sin(\cdots - \varphi_x \cdots)), \tag{12.10}$$

and

$$\frac{\partial(\cdots - \beta_l \cos(\varphi_x \cdots))}{\partial \varphi_x} = +\beta_l \sin(\varphi_x \cdots)), \tag{12.11}$$

which explains the correspondence. The equations of motions are satisfied in average (when inserted in the path integral). This is a consequence of the invariance under an arbitrary local shift for each integral (Schwinger–Dyson equations). When a shift $\delta\varphi_x$ is applied *after* the expansion in Fourier modes in the functional integral, we obtain insertions

$$e^{i\delta\varphi_x\left(\sum_\mu[-n_{x,\mu}+n_{x-\hat\mu,\mu}]\right)}. \tag{12.12}$$

As this is just a change of variable, it should not affect the result and for non-zero contributions, the sum of the indices in the exponent should be zero, which is precisely equation (12.5).

For the gauge indices, we can also assign in and out features to the plaquettes attached to a link in a way consistent with the Fourier expansion equation (7.32). For $\mu < \nu$, $m_{x,\mu,\nu}$ are in and $m_{x-\hat\nu,\mu,\nu}$ out, while for $\mu > \nu$, $m_{x,\nu,\mu}$ are out and $m_{x-\hat\nu,\nu,\mu}$ in. Again we can trace the signs by looking at the indices appearing with a specific $A_{x,\mu}$ integration using

$$e^{\beta_{pl.}\cos(A_{x,\mu}+A_{x+\hat\mu,\nu}-A_{x+\hat\nu,\mu}-A_{x,\nu})} = \sum_{m_{x,\mu,\nu}=-\infty}^{+\infty} e^{im_{x,\mu,\nu}(A_{x,\mu}+A_{x+\hat\mu,\nu}-A_{x+\hat\nu,\mu}-A_{x,\nu})}I_{m_{x,\mu,\nu}}(\beta_{pl.}).$$

When the other plaquette index ν is larger than μ, we have either

$$e^{im_{x,\mu,\nu}(A_{x,\mu}+\cdots)} \tag{12.13}$$

or

$$e^{im_{x-\hat\nu,\mu,\nu}(\cdots-A_{x-\hat\nu+\hat\nu,\mu}+\cdots)}, \tag{12.14}$$

which explains the signs in that case. When the other index is smaller, we interchange μ and ν and the signs.

In analogy with the continuum, we define the standard gauge-invariant lattice field strength tensor

$$f_{x,\mu,\nu} \equiv A_{x,\mu} + A_{x+\hat\mu,\nu} - A_{x+\hat\nu,\mu} - A_{x,\nu}. \tag{12.15}$$

With these notations,

$$\begin{aligned}
\frac{\partial S_{\text{CAHM}}}{\partial A_{x,\mu}} &= \beta_{\text{pl.}} \sum_{\nu>\mu}[\sin(f_{x,\mu,\nu}) - \sin(f_{x-\hat\nu,\mu,\nu})] \\
&\quad + \beta_{\text{pl.}} \sum_{\nu<\mu}[-\sin(f_{x,\nu,\mu}) + \sin(f_{x-\hat\nu,\nu,\mu})] \\
&\quad + \beta_{l.}\sin(d_{x,\mu}) \\
&= 0.
\end{aligned} \tag{12.16}$$

On the other hand, we have seen that the integration over $A_{x,\mu}$ yields the selection rule

$$\sum_{\nu > \mu} [m_{x,\mu,\nu} - m_{x-\hat{\nu},\mu,\nu}] + \sum_{\nu < \mu} [-m_{x,\nu,\mu} + m_{x-\hat{\nu},\nu,\mu}] + n_{x,\mu} = 0. \tag{12.17}$$

The geometric similarity between equation (12.16) with continuous variables and equation (12.17) with integer variables appears using equation (12.6) and the additional substitution

$$\beta_{\mathrm{pl.}} \sin(f_{x,\mu,\nu}) \leftrightarrow m_{x,\mu,\nu}. \tag{12.18}$$

As already discussed in section 7.3, these equations are a discrete versions of $\partial_\mu F^{\mu\nu} = J^\nu$ which implies $\partial_\mu J^\mu = 0$. In a similar way, equation (7.39) means that the link indices $n_{x,\mu}$ are determined by unrestricted plaquette indices $m_{x,\mu,\nu}$. We write this dependence as $n_{x,\mu}(\{m\})$ which is shorthand for equation (12.17).

12.2 A first look at topological solutions on the lattice

In section 12.1, we discussed the similarity between the continuous lattice equations of motion and the discrete tensor selection rules. In this subsection we discuss the effect of periodic boundary conditions on both sets of equations. We follow reference [1] and consider only the solvable cases: the $D = 1$ $O(2)$ spin model and the $D = 2$ pure gauge $U(1)$ model.

Topology is a branch of mathematics studying properties of geometrical objects that are invariant under smooth deformations. For instance, smooth maps of the circle into itself are characterized by winding numbers. In continuous field theory, topology provides useful tools to classify classical solutions (for a review see [2]). As an example for a smooth $U(1)$-valued field configuration in a two-dimensional theory with a circular boundary, the number of times the $U(1)$ field goes around its circle as we go around the boundary is an integer that cannot be modified by smooth deformations. However, with a lattice discretization, such considerations are only useful in the continuum limit.

For the $D = 1$ $O(2)$ spin model with PBC and N_τ sites, the equations of motion (12.3) with $A_{x,\mu} = 0$ are equivalent to the statement that $\sin(\varphi_{x+\hat{\imath}} - \varphi_x)$ takes the same value on every link. These equations have many solutions because they do not imply that $\varphi_{x+\hat{\imath}} - \varphi_x$ is constant ($\sin(\pi - \varphi_{x+\hat{\imath}} + \varphi_x) = \sin(\varphi_{x+\hat{\imath}} - \varphi_x)$). We will focus on the solutions that can be interpreted as continuous topological solutions in the continuum limit for PBC. If we impose that $\varphi_{x+\hat{\imath}} - \varphi_x$ is a small constant, we can obtain a solution that meets this requirement. Given any choice for the constant, we can then 'integrate' the equations: starting with some φ_0, we obtain φ_1, and so on until, due to PBC, we get an independent value for φ_0 which should be consistent with the initial value modulo an integer multiple of 2π. This follows from the field compactification. This approximately corresponds to a smooth mapping of the circle into itself provided that the successive changes can be made arbitrarily small. This can be accomplished by requiring that for all links

$$\varphi_{x+\hat{\imath}} - \varphi_x = \frac{2\pi}{N_\tau} \ell, \tag{12.19}$$

for a given integer ℓ. By taking N_τ large with fixed ℓ we obtain a solution which can be interpreted as a topological solution with winding number ℓ. In the limit $\ell \ll N_\tau$, these solutions have classical action

$$S_\ell = \beta\left(1 - \cos\left(\frac{2\pi}{N_\tau}\ell\right)\right)N_\tau \simeq \frac{\beta}{2}\left(\frac{2\pi}{N_\tau}\ell\right)^2 N_\tau. \tag{12.20}$$

We can calculate the quadratic fluctuations with respect to this solution using

$$1 - \cos\left(\frac{2\pi\ell}{N_\tau} + \delta\varphi_{x+1} - \delta\varphi_x\right) \simeq \frac{1}{2}\left(\frac{2\pi\ell}{N_\tau} + \delta\varphi_{x+1} - \delta\varphi_x\right)^2, \tag{12.21}$$

We can use the global $O(2)$ symmetry to set $\varphi_0 = 0$. Other values of φ_0 are taken into account by performing the integration over φ_0 which with our normalization of the measure yields a factor 1. By construction, the linear fluctuations vanish because the first derivatives are zero and all we need to calculate are the quadratic fluctuations

$$\Delta = \prod_{x=1}^{N_\tau-1} \int_{-\pi}^\pi \frac{d\varphi_x}{2\pi} e^{-S_\ell^{\text{quad.}}}, \tag{12.22}$$

with

$$S_\ell^{\text{quad.}} = \frac{\beta}{2}(\varphi_1^2 + (\varphi_2 - \varphi_1)^2 + \cdots + \varphi_{N_\tau-1}^2). \tag{12.23}$$

Following the standard quadratic path integral procedure from section 5.2, we find (see exercise below)

$$\Delta = N_\tau^{-1/2}(2\pi\beta)^{-(N_\tau-1)/2}. \tag{12.24}$$

We can now attempt to re-sum the topological contributions. This is delicate because we have assumed $\ell \ll N_\tau$, however, if β is large enough, the terms with large ℓ are exponentially suppressed. In the same spirit, we ignored the ℓ dependence of Δ. We can use the Poisson summation formula (see exercise below)

$$\sum_{\ell=-\infty}^\infty e^{-\frac{B}{2}\ell^2} = \sqrt{\frac{2\pi}{B}} \sum_{n=-\infty}^\infty e^{-\frac{(2\pi)^2}{2B}n^2}, \tag{12.25}$$

with $B = \beta(2\pi)^2/N_\tau$. Putting everything together, we get a semi-classical approximation of the partition function in the large β limit

$$Z \simeq (2\pi\beta)^{-N_\tau/2} \sum_{n=-\infty}^\infty \left(e^{-\frac{n^2}{2\beta}}\right)^{N_\tau}. \tag{12.26}$$

We now consider the solutions of the discrete equation (12.5). The solution is that $n_{x,1}$ should be constant. With PBC, and remembering that we use a classical action with a constant $(\beta(1 - \cos(\ldots)))$ in order to make the small fluctuations more obvious, we have the exact expression:

$$Z = \sum_{n=-\infty}^{\infty} (e^{-\beta} I_n(\beta))^{N_\tau}, \tag{12.27}$$

which can be compared to the semi-classical expression (12.26). Using the large β approximations

$$e^{-\beta} I_0(\beta) \simeq \frac{1}{\sqrt{2\pi\beta}} (1 + \mathcal{O}(1/\beta)), \tag{12.28}$$

and equation (7.22) in the same limit, we see the approximate correspondence between the two expressions.

Exercise 1: Using the results of section 5.2, check equation (12.24).

Solution. The Euclidean version of equation (5.16) is

$$\sqrt{\frac{m}{2\pi\hbar N_\tau \Delta\tau}} e^{-\frac{m(x_{N_\tau}-x_0)^2}{2\hbar N_\tau \Delta\tau}} = \left(\sqrt{\frac{m}{2\pi\hbar\Delta\tau}}\right)^{N_\tau} \prod_{i=1}^{N_\tau-1} \int_{-\infty}^{+\infty} dx_i e^{-\sum_{j=0}^{N_\tau-1} \frac{m}{2\hbar\Delta\tau}(x_{j+1}-x_j)^2}.$$

We can use this result with the substitution

$$\frac{m}{\hbar\Delta\tau} \to \beta,$$

in the exponential. Concerning the prefactors, we have $\left(\frac{1}{2\pi}\right)^{N_\tau-1}$ given that we have already performed the integration over φ_0. We don't have a prefactor corresponding to the one on the right-hand side of equation (5.16), so we need to multiply by $\left(\sqrt{\frac{\beta}{2\pi}}\right)^{N_\tau}$ in order to use equation (5.16) and divide by the same quantity since it is not there to start with. Using equation (5.16) with $x_0 = x_{N_\tau} = 0$, we get a factor $\sqrt{\frac{\beta}{2\pi N_\tau}}$. Putting everything together

$$\Delta = \left(\frac{1}{2\pi}\right)^{N_\tau-1} \left(\sqrt{\frac{2\pi}{\beta}}\right)^{N_\tau} \sqrt{\frac{\beta}{2\pi N_\tau}} = N_\tau^{-1/2}(2\pi\beta)^{-(N_\tau-1)/2}.$$

Exercise 2: Using equation (7.5), show that

$$\sum_{n=-\infty}^{+\infty} e^{i2\pi nx} = \sum_{n=-\infty}^{+\infty} \delta(x-n). \tag{12.29}$$

Using this result, derive the Poisson summation formula (12.25). Discuss the convergence of the sums in equation (12.25) numerically for $B = 0.1$, 1 and 10. Plot the truncated sums versions of equation (12.29) with

$$\sum_{n=-\infty}^{+\infty} \rightarrow \sum_{n=-n\max}^{+n\max} . \tag{12.30}$$

for $n\max = 1, 2, 3, 4$ and 30.

Solution. Equation (7.5) with $\varphi \rightarrow 2\pi x$ and the identity $\delta(2\pi x) = \frac{1}{2\pi}\delta(x)$ yields equation (12.29). We use this identity to rewrite the left-hand side of equation (12.25) and then interchange the sum and integral:

$$\sum_{\ell=-\infty}^{\infty} e^{-\frac{B}{2}\ell^2} = \int_{-\infty}^{+\infty} dx \sum_{\ell=-\infty}^{+\infty} \delta(x - \ell)e^{-\frac{B}{2}x^2},$$

$$= \int_{-\infty}^{+\infty} dx \sum_{n=-\infty}^{+\infty} e^{i2\pi n x}e^{-\frac{B}{2}x^2}, \tag{12.31}$$

$$= \sum_{n=-\infty}^{+\infty} \int_{-\infty}^{+\infty} dx e^{i2\pi n x}e^{-\frac{B}{2}x^2}.$$

Performing the Gaussian integrals as in section 5.2, we obtain the Poisson summation formula (12.25).

For the numerical values at $B = 0.1$, the sums converge to $7.926\ 654\ 595$ for $n\max = 20$ (lhs) and $n\max = 1$ (rhs); $B = 1$, $2.506\ 628\ 288$ for $n\max = 6$ (lhs) and $n\max = 1$ (rhs); $B = 10$, $1.013\ 475\ 898$ for $n\max = 1$ (lhs) and $n\max = 3$ (rhs); the details can be found in the appended Mathematica notebooks. In view of the exponential form of the terms of the sum it is clear that they are both convergent, and the numerical experiments conclusively show that the two sums are the same.

The convergence of the sum in equation (12.29) should be understood in integrals with test functions. If we look at the truncated versions of equation (12.29), we see that peaks at the integers build up slowly. The peak values are $2n\max + 1$. However, the oscillations in between the peaks remain significant. This is illustrated in figure 12.1 for $n\max = 1, 2, 3,$ and 4 and figure 12.2 for $n\max = 3, 6, 9,$ and 12. The oscillations in between the peaks remain of order one in magnitude. The feature persists for larger $n\max$ as illustrated in figure 12.3 for $n\max = 30$. However, as the $n\max$ increases, the oscillations are more rapid and in integrals with test functions, the only contributions come from the peaks at integer values.

12.3 Topology of $U(1)$ gauge theory and topological susceptibility

A similar construction can be carried for the $D = 2$ pure gauge $U(1)$ model with PBC. We consider a rectangular $N_s \times N_t$ lattice. The equation of motion requires that $\sin(f_{x,1,2})$ is constant. Following the analogy with the $O(2)$ case, we start with

$$f_{x,1,2} \equiv A_{x,1} + A_{x+\hat{1},2} - A_{x+\hat{2},1} - A_{x,2} = \delta, \tag{12.32}$$

with δ a constant to be determined with PBC. As seen in section 8.4 we can gauge transform the temporal links with a given spatial coordinate x_1 to the identity with

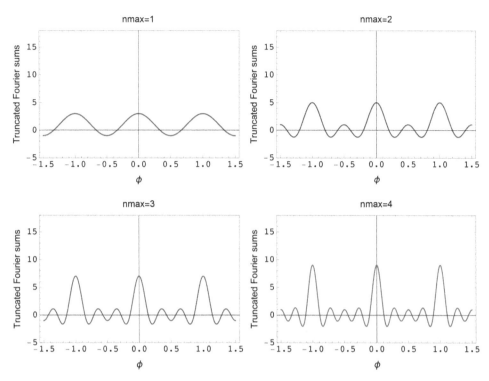

Figure 12.1. Truncations of equation (12.29) with $n$max = 1, 2, 3, and 4.

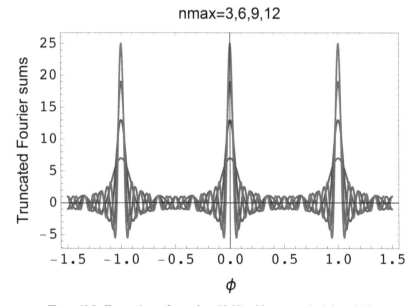

Figure 12.2. Truncations of equation (12.29) with $n$max = 3, 6, 9, and 12.

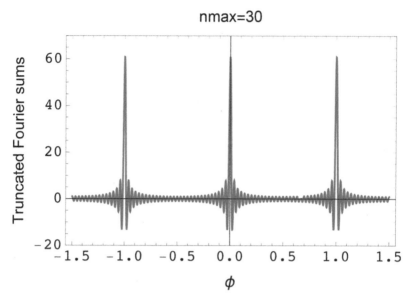

Figure 12.3. Truncation of equation (12.29) with $n$max $= 30$.

the exception of one time layer. We take this layer of nontrivial time links to be between $\tau = N_\tau - 1$ and N_τ which is identified with 0 due to the PBC. The space links with a fixed spatial coordinate, which can be visualized as a vertical ladder can be treated as the indices of a $D = 1$ $O(2)$ model changing by $-\delta$ at each step until we get to the 'last' rung and temporal links are present. The constancy of the 'last' plaquette requires that

$$A_{(x_1+1,N_\tau-1),2} - A_{(x_1,N_\tau-1),2} = N_\tau\delta. \tag{12.33}$$

Iterating in the spatial direction, we obtain PBC in the spatial direction provided that

$$\delta = \frac{2\pi}{N_s N_\tau}\ell. \tag{12.34}$$

The action for this topological solution is

$$S_\ell^{U(1)} \simeq \frac{\beta}{2}\left(\frac{2\pi}{N_s N_\tau}\ell\right)^2 N_s N_\tau. \tag{12.35}$$

We could have obtained another periodic solution by setting *all* the time links to 1 and imposing PBC in time for N_s independent $D = 1$ $O(2)$ models, but the action for these configurations is larger by a factor N_s^2.

The quadratic fluctuations can be calculated as in the $O(2)$ case but with extra complications due to the special time layer obtained after gauge-fixing. Keeping track of all the 2π factors and using Poisson summation for the winding numbers, we obtain the semi-classical approximation

$$Z_{U(1)} \simeq (2\pi\beta)^{-N_s N_\tau/2} \sum_{n=-\infty}^{\infty} (e^{-\frac{n^2}{2\beta}})^{N_s N_\tau}, \tag{12.36}$$

which agrees with the exact expression at leading order.

Exercise 3: Work out the details of these integrations following the $O(2)$ example. It is important to remove degrees of freedom associated with symmetries. There are $N_s N_\tau$ links. We can gauge fix $N_s(N_\tau - 1)$ temporal links and N_s remain in one layer of plaquettes connecting the N_s effective $O(2)$ models (on the other plaquette there are only two spatial links). Each of these N_s closed loops has a global $O(2)$ symmetry. The same symmetry also appears for the unintegrated temporal links. After getting rid of the redundant degrees of freedom we first integrate the spatial links keeping the unintegrated temporal links. The last step is the integration of the $N_s - 1$ temporal links. The exercise requires care and patience but it is satisfactory that again all the proper powers of β and 2π appear at the end.

In order to test the semi-classical picture we calculate the topological susceptibility. For this purpose we introduce

$$Z(\beta, \theta) = \prod_{x,\mu} \int_{-\pi}^{\pi} \frac{dA_{x,\mu}}{2\pi} e^{-S_{\text{gauge}} - i\theta Q}, \tag{12.37}$$

with the topological charge Q defined as

$$Q = \frac{1}{2\pi} \sum_x \sin(A_{x,1} + A_{x+\hat{1},2} - A_{x+\hat{2},1} - A_{x,2}). \tag{12.38}$$

The topological susceptibility is defined as

$$\chi = -\frac{d^2}{d\theta^2} \ln(Z)|_{\theta=0}. \tag{12.39}$$

It can be calculated using exact resummations [3]

$$Z(\beta, \theta) = \sum_{n=-\infty}^{\infty} \left[e^{-\beta} I_n \left(\sqrt{\beta^2 - \left(\frac{\theta}{2\pi}\right)^2} \right) \right.$$
$$\left. \times \left(\frac{\beta - \frac{\theta}{2\pi}}{\beta + \frac{\theta}{2\pi}} \right)^{n/2} \right]^{N_s N_\tau}. \tag{12.40}$$

If χ is dominated by configurations corresponding to winding number ± 1 where $|Q| \simeq 1$ in the continuum limit, it is plausible that in the large-β estimate, the topological susceptibility is almost saturated by the configurations with $Q = 0$ and ± 1

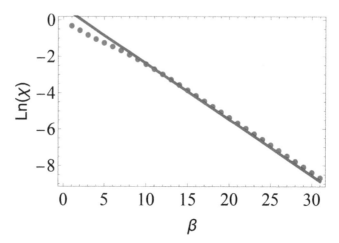

Figure 12.4. Logarithm of the topological susceptibility using the exact formula for $N_s = N_\tau = 8$ expanded up to order 5 (dots) and the semi-classical approximation equation (12.41) (continuous line). Reprinted with permission of APS from [1].

$$\chi \simeq (0)^2 1 + (1)^2 \exp\left(-\frac{\beta}{2}\frac{(2\pi)^2(1)^2}{N_s N_\tau}\right) + (-1)^2 \exp\left(-\frac{\beta}{2}\frac{(2\pi)^2}{N_s N_\tau}(-1)^2\right). \qquad (12.41)$$

Figure 12.4 shows that this estimate is reasonably good when β is large enough.

It is a common to misidentify the Fourier mode indices n in equation (12.27) as 'topological sectors'. They are rather 'rotor energy levels' $n^2/2$. The fact that Poisson summation interchanges these energy levels with the correctly identified topological sectors was observed in reference [4] in a version of the $O(2)$ model where the fluctuations are limited. Note also that it is possible to construct models where the approximate equations (12.26) and (12.36) are exact. The questions of topological configurations and duality are discussed for abelian gauge models of this type in various dimensions in references [5–9].

Exercise 4: Reproduce the results of figure 12.4. Discuss volume and mode truncation effects.

Solution. For $V = 64$, the sums up to $n$max $= 4$ appear to be sufficient for β up to 60. For $V = 32$, two more orders are necessary to accomplish the same result. See figure 12.5 and Mathematica notebooks at the end of the section.

Exercise 5: In the Fourier expansion of $e^{-\beta(1-\cos\phi)}$, replace the Bessel functions by an expression which coincide at leading order in $1/\beta$:

$$e^{-\beta}I_n(\beta) \to \frac{1}{\sqrt{2\pi\beta}}e^{-\frac{n^2}{2\beta}}. \qquad (12.42)$$

Using Poisson summation, calculate the modified series. Compare the approximations where the two kinds of sums (before and after Poisson summation) are

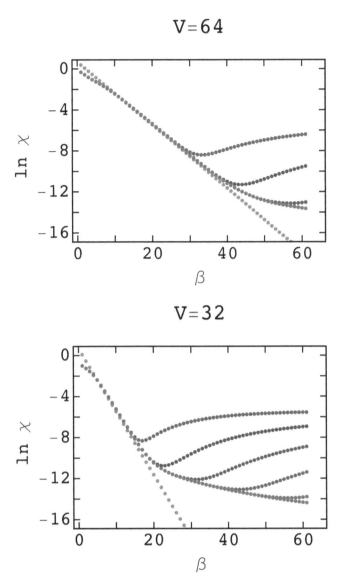

Figure 12.5. Logarithm of the topological susceptibility using the exact formula for $V = 64$ (top) and 32 (bottom) expanded up to order 2 (red), 3, 4 (purple), 5, 6 and 7 (blue) and the semi-classical approximation equation (12.41) (green dots forming a line). The increasing truncation orders go from top to bottom. Order 6 and 7 cannot be distinguished on the figure.

restricted to $-n\mathrm{max} \geqslant n \geqslant n\mathrm{max}$ modes for $\beta = 1$ and 10. Compare accurate Poisson summations with $e^{-\beta(1-\cos\phi)}$ for various β.

Solution. Using the procedure described in equation (12.31), we obtain

$$\frac{1}{\sqrt{2\pi\beta}} \sum_{n=-\infty}^{\infty} e^{-\frac{n^2}{2\beta}+in\varphi} = \sum_{\ell=-\infty}^{\infty} e^{-\frac{\beta}{2}(\varphi+2\pi\ell)^2}. \qquad (12.43)$$

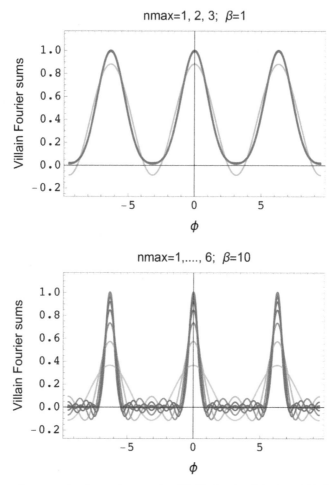

Figure 12.6. Successive approximations using equation (12.43), for $\beta = 1$ (top) and 10 (bottom). As $n$max increases, the color changes from green to blue.

This is a 'periodicized' Gaussian that behaves like $e^{-\beta(1-\cos\phi)}$ near 0, $\pm 2\pi$, …. This is called the 'Villain approximation' [10]. When β is not too small, the Gaussian centered around multiples of 2π have short tails and we only need a few terms to obtain accurate value say between -3π and 3π. For $\beta \geqslant 1$, $-3 \geqslant \ell \geqslant 3$ is sufficient in that range of φ. Concerning the Fourier expansion, for $\beta = 1$, $n$max = 3 is very close to the accurate answer. For $\beta = 10$, $n$max = 6 is very close to the accurate answer. This is illustrated in figure 12.6. The Villain approximation is very close to the original distribution at large β and very different at small β. For $\beta \simeq 1$, one can see small difference near the troughs as illustrated in figure 12.7. Arbitrary values of β can be explored with the Mathematica notebook appended. The use of the Villain approximation in Monte Carlo simulations has subtle aspects [11].

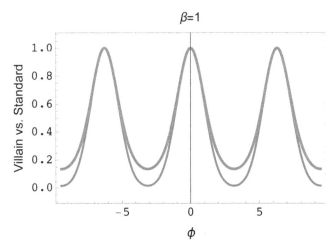

Figure 12.7. $e^{-\beta(1-\cos\phi)}$ (orange) and the Villain approximation (blue) for $\beta = 1$.

12.4 Mathematica notebooks

```
In[1]:= (*B is entered with a precision of 20 digits,
     the sums from both side of the Poisson summation
      formula are printed at succesive orders with 10 digits*)
     BB = SetPrecision[1/10, 20];
     Do[Print[nmax, "  ", SetPrecision[Sum[Exp[-BB*nn^2/2], {nn, -nmax, nmax}], 10],
        "  ", SetPrecision[Sqrt[2*Pi/BB] *
         Sum[Exp[-(2*Pi)^2*nn^2/(2*BB)], {nn, -nmax, nmax}], 10]], {nmax, 1, 22}];
```

```
1    2.902458849   7.926654595

2    4.539920355   7.926654595

3    5.815176658   7.926654595

4    6.713834587   7.926654595

5    7.286844180   7.926654595

6    7.617441957   7.926654595

7    7.790029130   7.926654595

8    7.871553538   7.926654595

9    7.906398287   7.926654595

10   7.919874181   7.926654595

11   7.924589905   7.926654595

12   7.926083077   7.926654595

13   7.926510877   7.926654595

14   7.926621781   7.926654595

15   7.926647795   7.926654595

16   7.926653317   7.926654595

17   7.926654377   7.926654595

18   7.926654562   7.926654595

19   7.926654590   7.926654595

20   7.926654595   7.926654595

21   7.926654595   7.926654595

22   7.926654595   7.926654595
```

```
In[3]:= BB = SetPrecision[1, 20];
     Do[Print[nmax, "  ", SetPrecision[Sum[Exp[-BB*nn^2/2], {nn, -nmax, nmax}], 10],
        "  ", SetPrecision[Sqrt[2*Pi/BB] *
         Sum[Exp[-(2*Pi)^2*nn^2/(2*BB)], {nn, -nmax, nmax}], 10]], {nmax, 1, 8}];
```

```
1    2.213061319   2.506628288

2    2.483731886   2.506628288

3    2.505949879   2.506628288

4    2.506620804   2.506628288

5    2.506628258   2.506628288

6    2.506628288   2.506628288

7    2.506628288   2.506628288

8    2.506628288   2.506628288
```

```
In[5]:= BB = SetPrecision[10, 20];
        Do[Print[nmax, "   ", SetPrecision[Sum[Exp[-BB * nn^2/2], {nn, -nmax, nmax}], 10],
           "   ", SetPrecision[Sqrt[2 * Pi / BB] *
              Sum[Exp[-(2 * Pi)^2 * nn^2 / (2 * BB)], {nn, -nmax, nmax}], 10]], {nmax, 1, 4}];
```

```
1   1.013475894   1.012885574

2   1.013475898   1.013475868

3   1.013475898   1.013475898

4   1.013475898   1.013475898
```

```
(*volume VV here 8x8, set of beta values, here 1 to 60 and naive formula*)
VV = 64; llmax = 60; pr = 20;
Do[bb[ll] = SetPrecision[1 + ll, pr];
   naiv[ll] = 2 * Exp[-(bb[ll] / 2) * (2 * Pi)^2 / VV], {ll, 0, llmax}];
t0 = Table[{bb[ll], Log[naiv[ll]]}, {ll, 0, llmax}];
li = ListPlot[Table[{bb[ll], Log[naiv[ll]]}, {ll, 0, llmax}],
   PlotStyle → {RGBColor[0, 0.5, 0], AbsolutePointSize[4]}]
```

Out[134]=

```
In[149]:= (*susceptibilities for nmax= 2, ...7*)
          Clear[theta, beta];
          eta = (1/2) * (beta - theta / (2 * Pi));
          etat = (1/2) * (beta + theta / (2 * Pi));
          Do[Z = Sum[(BesselI[nn, 2 * Sqrt[eta * etat]] * (Sqrt[eta / etat])^nn)^VV,
             {nn, -nnmax, nnmax}];
           der = (-D[Log[Z], {theta, 2}] /. {theta → 0});
           Do[susc[nnmax, ll] = der /. {beta → bb[ll]}, {ll, 0, llmax}];
           tt[nnmax] = Table[{bb[ll], Log[susc[nnmax, ll]]}, {ll, 0, llmax}], {nnmax, 2, 7}]
```

```
In[153]:=
```

```
(*convergence of  topological susceptibility at beta=50*)
TableForm[Table[{uu, susc[uu, 50]}, {uu, 2, 7}]]
```

Out[153]//TableForm=

2	0.00100951757130374
3	0.00002410695468241
4	$2.47647707950 \times 10^{-6}$
5	$2.36079724911 \times 10^{-6}$
6	$2.36063707351 \times 10^{-6}$
7	$2.36063701419 \times 10^{-6}$

```
(*putting everything together in a graph; tweaking the graph style *)
Do[lis[uu] = ListPlot[tt[uu],
    PlotStyle → {RGBColor[1 - uu / 7, 0, uu / 7], AbsolutePointSize[4]}], {uu, 2, 7}]
xmin = 0; xmax = 65;
ymin = -16; ymax = 0.;
dellx = 20; delly = 4;

xla = Join[Table[{x, x, {0.0, 0.}}, {x, xmin, xmax, dellx}],
Table[{x, " ", {0.025, 0.}, {GrayLevel[0.], AbsoluteThickness[1.2]}},
    {x, xmin - dellx, xmax + dellx, dellx / 2}]];

yla = Join[Table[{x, x, {0.0, 0.}}, {x, ymin, ymax, delly}],
Table[{x, " ", {0.025, 0.}, {GrayLevel[0.], AbsoluteThickness[1.2]}},
    {x, ymin - delly, ymax + delly, delly / 2}]];

xlat = Join[Table[{x, " ", {0.0, 0.}}, {x, xmin, xmax, dellx}],
Table[{x, " ", {0.025, 0.}, {GrayLevel[0.], AbsoluteThickness[1.2]}},
    {x, xmin - dellx, xmax + dellx, dellx / 2}]];

ylar = Join[Table[{x, " ", {0.0, 0.}}, {x, ymin, ymax, delly}],
Table[{x, " ", {0.025, 0.}, {GrayLevel[0.], AbsoluteThickness[1.2]}},
    {x, ymin - delly, ymax + delly, delly / 2}]];

final = Show[{li, Table[lis[uu], {uu, 2, 7}]}, Frame → True, AspectRatio → 0.7,
  Axes → True, PlotRange → {{xmin - 0.02, xmax + 0.02}, {ymin + 0.01, ymax - 0.01}},
  FrameTicks → {xla, yla, xlat, ylar},
  TextStyle → {FontSize → 20, FontWeight → "Plain", FontFamily → "Courier"},
  PlotLabel -> StyleForm["V=64"], FrameLabel → {StyleForm[β], StyleForm["ln χ "]},
  LabelStyle → Black, FrameStyle → Thickness[0.005]]
```

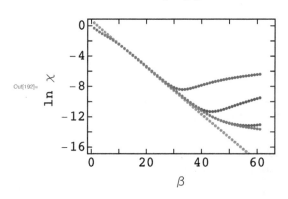

Out[192]=

In[194]:= `Export["mysusc64.pdf", final]`

Out[194]= mysusc64.pdf

In[438]:= (*pick beta and the range in phi*)
     beta = 10.; pmax = 3 * Pi;

In[439]:= (*the original Boltzmann weight*)
     bb[phi_] := Exp[-beta (1 - Cos[phi])]
     plbb = Plot[bb[phi], {phi, -3 * Pi, 3 * Pi},
        PlotRange → All, PlotStyle → {RGBColor[0, 0.5, 1], Thickness[0.01]}]

Out[440]=

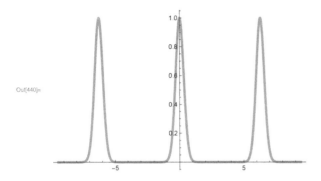

In[441]:=

     (*truncated sums of the Villain approximation *)
     PG[phi_, nmax_] :=
      Sum[Exp[-Min[{(beta / 2) * (phi - 2 * Pi * nn) ^ 2, 100}]], {nn, -nmax, nmax}];
     nmax = 5;

In[442]:= (*plotting for nmax=0,1,2, and 3 *)
     Do[Print[Plot[PG[x, m], {x, -pmax, pmax},
          PlotStyle → {RGBColor[1 - m / nmax, 0, m / nmax]}, PlotRange → All]], {m, 0, nmax}];

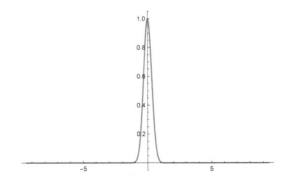

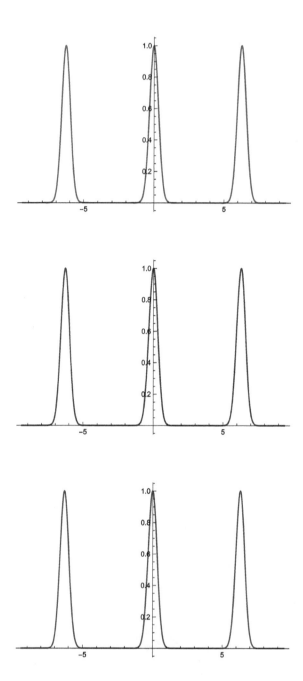

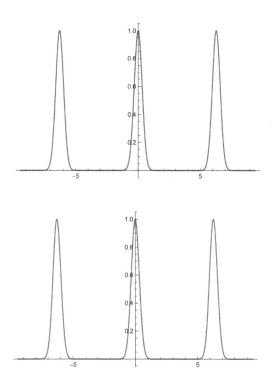

In[443]:= (*difference between nmax=1 and 3 on a log scale*)
Plot[Log[Abs[PG[x, 1] - PG[x, 3]]], {x, -pmax, pmax}, PlotRange → All]

Out[443]=

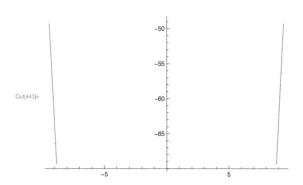

In[444]:= (*comparing Villain with original*)
Plot[{bb[phi], PG[phi, 3]}, {phi, -pmax, pmax}, PlotRange → All]

Out[444]=

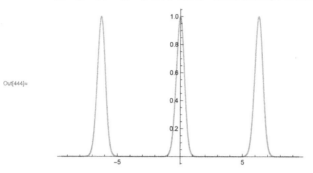

In[445]:= Plot[bb[phi] - PG[phi, 3], {phi, -pmax, pmax}, PlotRange → All]

Out[445]=

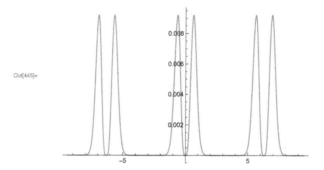

In[446]:= plvv = Plot[PG[phi, 3], {phi, -3 * Pi, 3 * Pi}, PlotRange → All,
PlotStyle → {RGBColor[0, 0.4, 0.9], Thickness[0.008]}]

Out[446]=

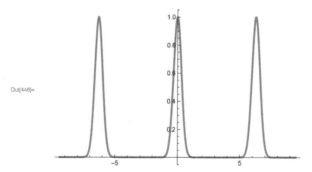

In[447]:=

```
(*succesive Fourier truncations for Villain: nmax=1, ..., 6*)
TV[phi_, nmax_] := (1 / Sqrt[2 * Pi * beta]) *
    (1 + Sum[2 * Cos[nn * phi] * Exp[-nn^2 / (2 * beta)], {nn, 1, nmax}])
nmaxmax = 6;
Do[Print[Show[plvv, plg[nmax] = Plot[TV[phi, nmax], {phi, -pmax, pmax},
      PlotStyle → {RGBColor[0, 1 - nmax / nmaxmax, nmax / nmaxmax]},
      PlotRange → {All, {-0.4, 1.1}}],
    BaseStyle → {FontFamily → "Courier", FontSize → 14},
    Frame → True, FrameLabel → {Style[Text[ϕ], FontSize → 14],
      Style[Text["Truncated Fourier sums"], FontSize → 14],
      Style[Text["nmax=" <> ToString[nmax]], FontSize → 14]},
    LabelStyle → Black]], {nmax, 1, nmaxmax}]
```

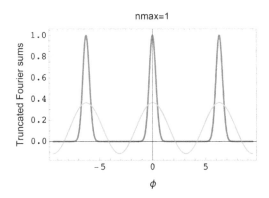

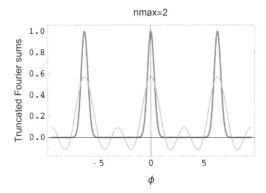

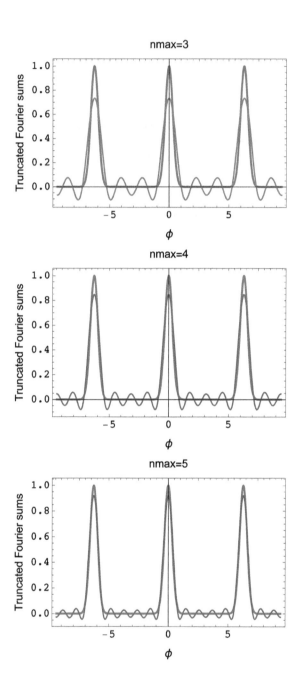

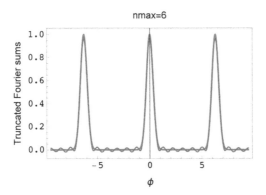

In[450]:=

```
final = Show[plvv, Table[plg[m], {m, 1, nmaxmax}],
  PlotStyle → {RGBColor[1 - nmax / nmaxmax, 0, nmax / nmaxmax]}, PlotRange →
    {All, {-0.2, 1.1}}, (*BaseStyle→{FontFamily→"Courier",FontSize→14},*)
  TextStyle → {FontSize → 14, FontWeight → "Plain", FontFamily → "Courier"},
  Frame → True, FrameLabel → {Style[Text[ϕ], FontSize → 14],
    Style[Text["Villain Fourier sums"], FontSize → 14],
    Style[Text["nmax=1,...., 6;  β=10"], FontSize → 14]}, LabelStyle → Black]
```

Out[450]:=

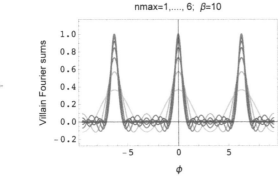

In[451]:= `SetDirectory[]`

Out[451]= /Users/yannickmeurice

In[452]:= `SetDirectory["Dropbox/amodman/chapter12"]`

Out[452]= /Users/yannickmeurice/Dropbox/amodman/chapter12

In[113]:= **FileNames[]**

Out[ ]= {64suscgraph_Part1.pdf, 64suscgraph_Part2.pdf, 64suscgraph_Part3.pdf,
64suscgraph.pdf, advanced.log, advanced.tex, beta10villain.pdf, .DS_Store,
final, mysusc32.pdf, mysusc64.pdf, poissonconvergence_Part1.pdf,
poissonconvergence_Part2.pdf, publicsoftware, sum1.pdf, sum2.pdf, sum30.pdf,
sum36912.pdf, sum3.pdf, sum4.pdf, susc.pdf, temp, villain10.pdf, villain1.pdf}

In[453]:= **Export["villain10.pdf", final]**

Out[453]= villain10.pdf

    "villain1.pdf"

In[454]:= **(*bessel expansions*)**
TI[phi_, nmax_] := Exp[-beta] *
    (BesselI[0, beta] + Sum[2 * Cos[nn * phi] * BesselI[nn, beta], {nn, 1, nmax}])

In[116]:= **TI[phi, 3]**

In[455]:= 0.00004539992976248485400 (2815.7166284662530 + 5341.9766074025070 Cos[phi] +
4563.0379354520050 Cos[2 phi] + 3516.7614332217050 Cos[3 phi])

Out[455]= 0.0000453999 (2815.72 + 5341.98 Cos[phi] + 4563.04 Cos[2 phi] + 3516.76 Cos[3 phi])

In[456]:= **(*successive Bessel orders*)**
nmaxmax = 6;
Do[Print[Show[plbb, pl[nmax] = Plot[TI[phi, nmax], {phi, -pmax, pmax},
    PlotStyle → {RGBColor[1 - nmax / nmaxmax, 0, nmax / nmaxmax]},
    PlotRange → {All, {-0.4, 1.1}}],
  BaseStyle → {FontFamily → "Courier", FontSize → 14},
  Frame → True, FrameLabel → {Style[Text[φ], FontSize → 14],
    Style[Text["Truncated Fourier sums"], FontSize → 14],
    Style[Text["nmax=" <> ToString[nmax]], FontSize → 14]},
  LabelStyle → Black]], {nmax, 1, nmaxmax}]

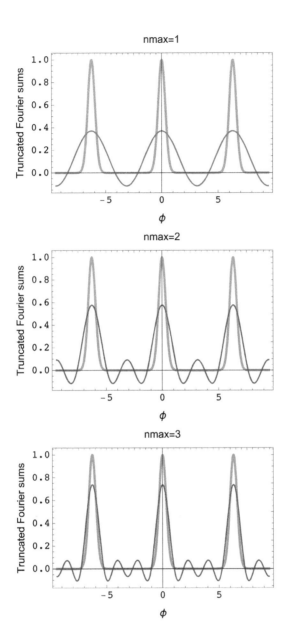

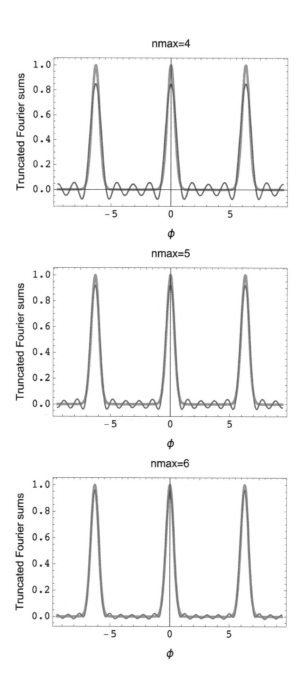

12.5 Large field effects in perturbation theory

As we discussed before, perturbation theory has played an essential role in developing and establishing the standard model of electroweak and strong inter-actions. The renormalizability of the theory guarantees that it is meaningful to calculate the radiative corrections at any order in perturbation theory. On the other hand, a generalization of Dyson's argument [12] (what happens for instance if in a $\lambda\phi^4$ theory we take $\lambda < 0$) or estimates of the number of diagrams suggest that the perturbative series are divergent and one needs to truncate the series in order to get a finite answer for a nonzero coupling. In the absence of a definite prescription to deal with this problem, one usually relies on the 'rule of thumb' which consists in dropping the smallest contribution at a given coupling and all the higher order terms. Clearly, this procedure has a limited accuracy and it is not always obvious how to estimate the error or to decide if one needs to calculate one extra order.

The problem is particularly acute for QCD corrections because they are large even at low order. Next to leading order corrections are important [13] for multijet processes to be studied experimentally with the LHC. Another example [14] is the hadronic width of the Z^0 where the term of order α_s^3 is more than 60 percent of the term of order α_s^2 and contributes to one part in 1000 of the total width (a typical experimental error for individual LEP experiments). It is not clear that the next term would improve the accuracy of the calculation. These calculations are time consuming and one would like to address the lack of convergence of perturbative series in a systematic way.

The factorial growth of perturbative series is related to configurations with arbitrary large fields. Large order coefficients are dominated by large field config-urations for which the expansion of the exponential of the perturbation at a given order is not a good approximation. In examples [15] where the modified coefficients can be calculated numerically, it was shown that a large field cutoff drastically affects the asymptotic behavior of the perturbative series. This is summarized in figure 12.8.

It is possible to understand the basic idea of the divergence of perturbative with a simple integral. The lack of uniform convergence means that

$$\int_{-\infty}^{+\infty} d\phi e^{-\frac{1}{2}\phi^2 - \lambda\phi^4} \neq \sum_{0}^{\infty} \frac{(-\lambda)^l}{l!} \int_{-\infty}^{+\infty} d\phi e^{-\frac{1}{2}\phi^2} \phi^{4l}. \tag{12.44}$$

The peak of the integrand of the rhs moves too fast when the order increases. At order ℓ in ϕ^4, we have the integrand

$$\phi^{4\ell} e^{-\frac{1}{2}\phi^2}, \tag{12.45}$$

which has an extremum when $-\phi + 4\ell/\phi = 0$ and the peak of the integrand at order ℓ is at $\phi^2 = 4\ell$. The truncation of $\exp(-\lambda\phi^4)$ at order ℓ is a good approximation if $\lambda\phi^4 \ll \ell$. But at the peak, $\lambda\phi^4 = \lambda 16\ell^2$ and we need $\lambda 16\ell^2 \ll \ell$. Consequently, as we increase the order, the region of validity become smaller $\lambda \ll 1/(16\ell)$. These series are called asymptotic series.

On the other hand, if we introduce a field cutoff, the peak moves outside of the integration range and

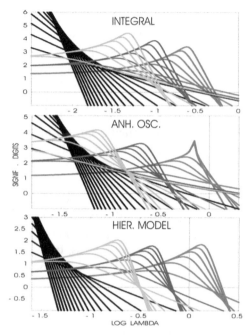

Figure 12.8. Significant digits obtained with regular perturbation theory at odd orders (black) and with $\phi_{max} = 3$ (green), 2.5 (blue) and 2 (red), for the integral, the ground state energy of the anharmonic oscillator and the renormalized mass of Dyson's model. (Reprinted with permission of APS from [15] which provides the details.)

$$\int_{-\phi_{max}}^{+\phi_{max}} d\phi e^{-\frac{1}{2}\phi^2 - \lambda\phi^4} = \sum_0^\infty \frac{(-\lambda)^l}{l!} \int_{-\phi_{max}}^{+\phi_{max}} d\phi e^{-\frac{1}{2}\phi^2} \phi^{4l}. \tag{12.46}$$

We expect that for a finite lattice, the partition function Z calculated with a field cutoff is convergent and that $\ln(Z)$ has a finite radius of convergence controlled by the zeros of the partition function in the complex coupling plane. For a given maximal order K and coupling λ, we can adjust $\phi_{max}(\lambda, K)$ in order to minimize the error. The strong coupling can be used to calculate approximately this optimal $\phi_{max}(\lambda, K)$ [16]. The perturbative 'coefficients' $a_k(\phi_{max}(\lambda, K))$ are then depending on λ.

The large field behavior is often related to classical solutions of the classical equations of motion at Euclidean time. For the quantum mechanical problem of one degree of freedom with a double-well potential $V(q) = (1/2)q^2 - gq^3 + (g^2/2)q^4$, we have solutions called 'instantons' which are nicely introduced in S Coleman's Erice lectures 'The uses of instantons' [17], where more references can be found. The difference between the perturbative series for the ground state and the numerical result is accounted by the one-instanton effect:

$$\Delta E_0 = -(g\pi)^{-1/2} e^{-\frac{1}{6g^2}}. \tag{12.47}$$

DOUBLE WELL

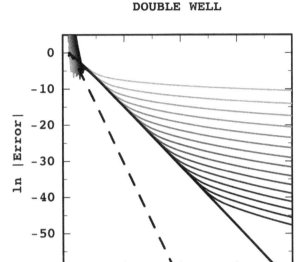

Figure 12.9. $\ln(|\text{Error}|)$ for order 1–15 (in g^2) versus $1/g^2$ for the ground state of the double-well potential. As the order increases, the curves get darker. The thicker dark curve is $\ln((g\pi)^{-1/2}e^{-\frac{1}{6g^2}})$ (1-instanton). The dash curve is $\ln(\sqrt{6/\pi}g^{-1}e^{-\frac{1}{3g^2}})$. Reprinted with permission of APS from [18] which provides full explanations.

This is illustrated in figure 12.9. If we use the rule of thumb with the series we find an estimate of the perturbative error which is about the square of the one-instanton effect. This is illustrated in figure 12.9 and explained in reference [18].

12.6 Remarks about the strong coupling expansion

The character expansions used in chapter 7 were initially developed in the context of the strong coupling (also called high-temperature) expansion [19, 20]. This expansion can be constructed by path enumerations and converges when β is small enough. We would like to point out that the path representation is exact and remains useful for any β even when $\beta > \beta_c$ and the series in powers of β does not converge.

For the $D = 2$ Ising model, the path representation of the partition function reads

$$Z = 2^V (\cosh(\beta))^{2V} \sum_{N_b} (\tanh(\beta))^{N_b} \mathcal{N}(N_b). \tag{12.48}$$

The sum is over closed paths with N_b links. At each site, there are either 0, 2 or 4 links belonging to the path. Each links costs a factor $\tanh(\beta)$. It is clear that the tensor assembly reproduces these paths. We can then calculate the average energy

$$\langle E \rangle = -\frac{\partial}{\partial \beta} \ln Z = -\tanh(\beta) \left(2V + \frac{\langle N_b \rangle}{\sinh^2(\beta)} \right). \tag{12.49}$$

This representation is exact and can be used for arbitrarily large β. In the limit of large β, we find that

$$\frac{\langle E \rangle}{V} = -2 + \mathcal{O}(\exp(-8\beta)). \tag{12.50}$$

The correction to -2 comes from the flip of a single spin in a uniform configuration of 1 or -1. Consequently,

$$\frac{\langle N_b \rangle}{V} = 1 + \mathcal{O}(\exp(-2\beta)), \text{ when } \beta \to \infty. \tag{12.51}$$

In other words, when β is very large, half of the links are occupied in average and the paths are densely filling the surface. These paths can be sampled with the worm algorithm [21]. Samples for $\beta = 0.4$, 0.5, 0.7 and 1 are shown in figure 12.10. It is interesting to observe the change of behavior as we go across $\beta_c = 0.440\ 7\ldots$. For $\beta = 0.4$, there are small connected parts of order $\tanh(\beta)^4$ and $\tanh(\beta)^6$. As we cross β_c, long paths develop signaling the breakdown of the strong coupling expansion. However, the worm sampling remains effective. The paths can also be constructed with blocked tensors [22].

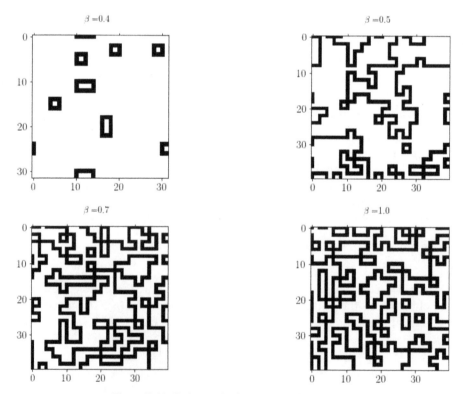

Figure 12.10. Path samples for $\beta = 0.4$, 0.5, 0.7 and 1.

References

[1] Meurice Y 2020 *Phys. Rev.* D **102** 014506

[2] Rajaraman R 1982 *Solitons and Instantons. An Introduction to Solitons and Instantons in Quantum Field Theory* (Amsterdam: North-Holland)

[3] Gattringer C, Kloiber T and Müller-Preussker M 2015 *Phys. Rev.* D **92** 114508

[4] Akerlund O and de Forcrand P 2015 *J. High Energy Phys.* **06** 183

[5] Savit R 1977 *Phys. Rev. Lett.* **39** 55–8

[6] Banks T, Myerson R and Kogut J B 1977 *Nucl. Phys.* B **129** 493–510

[7] Savit R 1980 *Rev. Mod. Phys.* **52** 453–87

[8] Gattringer C, Goschl D and Sulejmanpasic T 2018 *Nucl. Phys.* B **935** 344–64

[9] Sulejmanpasic T and Gattringer C 2019 *Nucl. Phys.* B **943** 114616

[10] Villain J 1975 *J. Phys. (France)* **36** 581–90

[11] Janke W and Kleinert H 1986 *Nucl. Phys.* B **270** 135–53

[12] Dyson F J 1952 *Phys. Rev.* **85** 631–2

[13] Gianotti F and Mangano M L 2005 LHC physics: The First one–two year(s) *2nd Italian Workshop on the Physics of Atlas and CMS* pp 3–26

[14] Larin S, van Ritbergen T and Vermaseren J 1994 *Phys. Lett.* B **320** 159–64

[15] Meurice Y 2002 *Phys. Rev. Lett.* **88** 141601

[16] Kessler B, Li L and Meurice Y 2004 *Phys. Rev.* D **69** 045014

[17] Coleman S 1985 *Aspects of Symmetry: Selected Erice Lectures* (Cambridge: Cambridge University Press)

[18] Meurice Y 2006 *Phys. Rev.* D **74** 096005

[19] Balian R, Drouffe J M and Itzykson C 1975 *Phys. Rev.* D **11** 2104–19

[20] Itzykson C and Drouffe J 1991 *Statistical Field Theory: Volume 1, From Brownian Motion to Renormalization and Lattice Gauge Theory* (Cambridge Monographs on Mathematical Physics) (Cambridge: Cambridge University Press)

[21] Prokof'ev N and Svistunov B 2001 *Phys. Rev. Lett.* **87** 160601

[22] Foreman S, Giedt J, Meurice Y and Unmuth-Yockey J 2018 *Phys. Rev.* E **98** 052129

IOP Publishing

Quantum Field Theory
A quantum computation approach
Yannick Meurice

Appendix A

Elements of group theory and Lie algebras

This appendix presents a few practical results needed for field theory. It is by no means a systematic discussion.

We say that $(G, \cdot)$ is a group (denoted multiplicatively) if for arbitrary group elements $g, g', g'', \ldots$:

(i) $g \cdot g' \in G$ (closure);
(ii) $(g \cdot g') \cdot g'' = g \cdot (g' \cdot g'')$ (associativity);
(iii) there is a unique identity element e such that $e \cdot g = g \cdot e = g$;
(iv) each element g has a unique inverse g^{-1} such that $g \cdot g^{-1} = e = g^{-1} \cdot g$.

The requirement that the left inverse g_L^{-1} and right inverse g_R^{-1} are the same group element is actually superfluous because it is a consequence of (ii) and (iii) : $g_L^{-1} g g_R^{-1} = e g_R^{-1} = g_L^{-1} e$.

In the context of field theory, the groups of transformations that leave specific quadratic forms invariant play an important role. The main examples are the groups $O(N, M)$, which leave the real quadratic form

$$\sum_{i=1}^{N} x_i^2 - \sum_{i=N+1}^{N+M} x_i^2, \tag{A.1}$$

invariant and $U(N, M)$, which leave the complex quadratic form

$$\sum_{i=1}^{N} |z_i|^2 - \sum_{i=N+1}^{N+M} |z_i|^2, \tag{A.2}$$

invariant. If in addition, we require that the linear transformation has a determinant 1, we restrict these groups to the special subgroups $SO(N, M)$ and $SU(N, M)$. As we will see, group elements can be obtained by exponentiating generators. It is useful to know the matrix identity

$$\det(C) = \exp(\mathrm{Tr}(\ln(C))). \tag{A.3}$$

doi:10.1088/978-0-7503-2187-7ch13

We now discuss in more detail the case $M = 0$. The groups are denoted $SO(N)$ (orthogonal) and $SU(N)$ (unitary). The irreducible unitary representations of these groups are finite dimensional, known and tabulated. The behavior of the group elements near the identity can be expressed using small continuous parameters (infinitesimal).

An element of $SO(N)$ is a $N \times N$ real matrix O, such that determinant$(O) = 1$ and $OO^T = 1$. If we write $O = 1 + \epsilon A$, then at first order in ϵ (neglecting order ϵ^2), $OO^T \simeq 1 + \epsilon(A + A^T)$. The orthogonality requirement is satisfied if $A^T = -A$, in other words if A is antisymmetric. The matrix A is called an *infinitesimal generator* of the orthogonal group. We can write an arbitrary one by using linear combinations of the $N(N-1)/2$ matrices with a single 1 above the diagonal, a single -1 at the transposed position, all the other matrix elements being zero. If we multiply A by i, we obtain an Hermitian matrix.

A element of $SU(N)$ is a $N \times N$ complex matrix U such that determinant$(U) = 1$ and $UU^\dagger = 1$. If we write $U = 1 + i\epsilon H$, then at first order in ϵ, $UU^\dagger \simeq 1 + i\epsilon(H - H^\dagger)$. The unitarity requirement is satisfied if $H^\dagger = H$, in other words if H is Hermitian. The matrix H is called an *infinitesimal generator* of the unitary group.

The commutator of two Hermitian matrices is an anti-Hermitian matrix which we can write as i times an Hermitian matrix. More generally, the generators of Lie groups (groups like $SU(N)$ or $SO(N)$ which depends smoothly on the parameters used to write the group elements) form a Lie algebra defined by

$$[T^A, T^B] = if^{ABC}T^C. \tag{A.4}$$

The f^{ABC} are called the structure constants and all the possible forms have been classified by Lie and Cartan.

IOP Publishing

Quantum Field Theory
A quantum computation approach
Yannick Meurice

Appendix B

Spaces defined by quadratic forms

In this appendix, we discuss a special class of curved spaces that have a constant curvature. For a more detailed discussion, see [1]. We use curvilinear coordinates

$$x^\mu, \mu = 1, 2, \ldots n, \tag{B.1}$$

embedded in flat space with $n + 1$ coordinates

$$X^A, A = 1, 2, \ldots n, n + 1 \tag{B.2}$$

and a constant metric η_{AB} that has only ± 1 elements.

We first consider the n-dimensional sphere S_n, which is important to understand dimensional regularization. We use a flat metric η_{AB} with completely positive signature $(+, +, \ldots, +)$. A sphere of radius R is defined by

$$X^A X^B \eta_{AB} = (X^1)^2 + (X^2)^2 + \cdots + (X^{n+1})^2 = R^2. \tag{B.3}$$

For $n = 1$ the solutions can be written with polar coordinates

$$X^1 = R \cos \phi, \tag{B.4}$$

$$X^2 = R \sin \phi, \tag{B.5}$$

with $0 \leqslant \phi \leqslant 2\pi$. For $n \geqslant 2$, we set

$$X^{n+1} = R \cos \theta_{n-1}. \tag{B.6}$$

For $0 < \theta_{n-1} < \pi$, we have a $n - 1$-dimensional sphere of radius $R \sin \theta_{n-1}$. At $\theta = 0$ or π, the sphere shrinks into a point. For $n = 2$, it is easy to visualize the collection of circles shrinking into a point at the poles. For $n > 2$, we can iterate the process.

We now turn to the de Sitter space (dS_n) which has signature $(+, +, \ldots, +, -)$. The space is defined as the solutions of

$$X^A X^B \eta_{AB} = (X^1)^2 + (X^2)^2 + \cdots + (X^n)^2 - (X^{n+1})^2 = R^2. \tag{B.7}$$

doi:10.1088/978-0-7503-2187-7ch14

We treat X^{n+1} as the 'time'. For $n = 1$, we have two space-like hyperbolas:

$$X^1 = \pm R \cosh \chi, \tag{B.8}$$

$$X^2 = R \sinh \chi, \tag{B.9}$$

with $\chi \in \mathbb{R}$. For $n \geqslant 2$, we set

$$X^{n+1} = R \sinh \chi. \tag{B.10}$$

again with $\chi \in \mathbb{R}$. And for each χ, we have a S_{n-1} defined by

$$(X^1)^2 + (X^2)^2 + \cdots + (X^n)^2 = (R \cosh \chi)^2 \geqslant R^2. \tag{B.11}$$

For $n = 2$, we have a one-sheet space-like hyperboloid. Topologically, dS_n is equivalent to $S_{n-1} \times \mathbb{R}$.

We can change the sign in front of R^2 in dS_n and obtain the hyperbolic spaces $\mathbb{H}_n$:

$$X^A X^B \eta_{AB} = (X^1)^2 + (X^2)^2 + \cdots + (X^n)^2 - (X^{n+1})^2 = -R^2. \tag{B.12}$$

For $n = 1$, we have two space-like hyperbolas. It is just like the de Sitter case with X^1 and X^2 interchanged. For $n \geqslant 2$, we set

$$X^{n+1} = \pm R \cosh \chi, \tag{B.13}$$

with $\chi \geqslant 0$. And for each χ, we have a S_{n-1} defined by

$$(X^1)^2 + (X^2)^2 + \cdots + (X^n)^2 = (R \sinh \chi)^2, \tag{B.14}$$

and we have a collection of sphere shrinking to a point when $\chi \to 0$. For $n = 2$, we have two time-like hyperboloids (past and future). Topologically, each of the two components of $\mathbb{H}_n$ is equivalent to $\mathbb{R}^n$.

Finally, we discuss the Anti-de Sitter (AdS_n) spaces obtained by flipping the sign of $(X^n)^2$ in $\mathbb{H}_n$

$$X^A X^B \eta_{AB} = (X^1)^2 + (X^2)^2 + \cdots - (X^n)^2 - (X^{n+1})^2 = -R^2. \tag{B.15}$$

For $n = 2$, this is equivalent to dS_n with the order of the variables flipped. For $n \geqslant 3$, we set

$$(X^n)^2 + (X^{n+1})^2 = (R \cosh \chi)^2, \tag{B.16}$$

with $\chi \geqslant 0$. For each χ this is a S_1. And for each χ, we have a S_{n-1} defined by

$$(X^1)^2 + (X^2)^2 + \cdots + (X^{n-1})^2 = (R \sinh \chi)^2, \tag{B.17}$$

and we have a collection of S_{n-2} shrinking to a point when $\chi \to 0$. This collection is equivalent to $\mathbb{R}^{n-1}$. Topologically, AdS_n is equivalent to $\mathbb{R}^{n-1} \times S_1$.

An interesting example is AdS_3. The following set of coordinates

$$
\begin{aligned}
X^1 &= R \sinh\chi \cos\phi \\
X^2 &= R \sinh\chi \sin\phi \\
X^3 &= R \cosh\chi \cos\tau \\
X^4 &= R \sinh\chi \sin\tau,
\end{aligned}
\tag{B.18}
$$

with $\chi \geqslant 0$, $0 \geqslant \phi \geqslant 2\pi$, and $0 \geqslant \tau \geqslant 2\pi$ covers the entire space once. These are called the global coordinates. The reason we have $\chi \geqslant 0$ is that changing the sign of χ has the same effect as $\phi \to \phi + \pi$.

Exercise 1: Calculate the metric of AdS_3 in global coordinates.

Solution. $g_{\chi\chi} = 1$, $g_{\phi\phi} = \sinh\chi^2$, $g_{\tau\tau} = -\cosh\chi^2$, the others are zero.

It is also possible to calculate the geodesics for the spaces discussed above. A first observation is that the quantity that when varied produces the geodesic equation in section 2.9, can be written as

$$
\tilde{L} \equiv g_{\mu\nu}\dot{x}^\mu \dot{x}^\nu = \eta_{AB}\dot{X}^A \dot{X}^B.
\tag{B.19}
$$

This is nice because the metric is now constant, however, in order to impose the condition

$$
X^A X^B \eta_{AB} = C,
\tag{B.20}
$$

for a constant C, we need to add a Lagrange multiplier. We now consider

$$
\tilde{L}_\lambda \equiv \eta_{AB}\dot{X}^A \dot{X}^B + \lambda(X^A X^B \eta_{AB} - C).
\tag{B.21}
$$

The Euler–Lagrange equations become linear:

$$
\ddot{X}^B = \lambda X^B.
\tag{B.22}
$$

Furthermore, the constancy of $X^A X^B \eta_{AB}$ along the geodesic requires

$$
X^A \dot{X}^B \eta_{AB} = 0.
\tag{B.23}
$$

We have $n(n-1)/2$ constants of motion

$$
L^{AB} \equiv X^A \dot{X}^B - \dot{X}^A X^B,
\tag{B.24}
$$

for $A \neq B$. $\dot{L}^{AB} = 0$ is a consequence of the geodesic equation (B.22) and the antisymmetry in AB. If η_{AB} has m minuses and $n - m + 1$ *pluses* $\tilde{L}_\lambda$ is invariant under $O(n - m + 1, m)$ transformations. This group has $n(n + 1)/2$ generators which correspond to the constants of motion using Noether's theorem. In addition $\eta_{AB}\dot{X}^A \dot{X}^B$ is also a constant of motion as we can see by taking the derivative with respect to s, using the geodesic equation (B.22) and the requirement (B.23). This can also be seen from the identity

$$L^{AB}L_{AB} = 2C\dot{X}^A\dot{X}^B\eta_{AB}, \tag{B.25}$$

which is a constant.

For arbitrary $n \geqslant 2$, we have

$$\epsilon_{A_1A_2...A_{n+1}}L^{A_1A_2}X^{A_3} = 0. \tag{B.26}$$

If the initial position and velocity are not collinear, this relation selects a plane defined by the initial position and velocity and including the origin. This is easy to visualize for $n = 2$ and remains true in any higher dimension. The intersection of this plane with the n-dimensional embedded curved space defines a one-dimensional object which is the geodesic curve. From this reasoning, we see that for S_n, the geodesics are great circles. For H_2, the geodesics are hyperbolas at the intersection of planes including the origin and time-like hyperboloids. For dS_2, the intersection of the $X^3 = 0$ plane with the dS_2 space-like hyperboloid is a circle. If we tilt this plane continuously into the $X^2 = 0$ plane, we end up with a hyperbola. At an intermediate angle of 45 degrees, two lines appear. With the sign convention used in this appendix, these lines are $(\pm R, s, s)$ and the velocities are light-like. This is illustrated in figures 2.1, 2.3–2.5.

Circles and hyperbolas appear in the general solution of the geodesic equation (B.22) for $\lambda \neq 0$:

$$X^A(s) = X_+^A e^{\sqrt{\lambda}s} + X_-^A e^{-\sqrt{\lambda}s}. \tag{B.27}$$

The constants are related to the initial position and velocity:

$$X_{\pm}^A = \frac{1}{2}(X^A(0) \pm \frac{1}{\sqrt{\lambda}}\dot{X}^A(0)). \tag{B.28}$$

Imposing the conditions (B.20) and (B.23), we obtain

$$\eta_{AB}X_+^A X_+^B = \eta_{AB}X_-^A X_-^B = 0, \quad 2\eta_{AB}X_+^A X_-^B = C, \tag{B.29}$$

and

$$\lambda = -\frac{\eta_{AB}\dot{X}^A\dot{X}^B}{C}. \tag{B.30}$$

For S_n, $C > 0$ and $\eta_{AB}\dot{X}^A\dot{X}^B > 0$, consequently $\lambda < 0$ and the solution involves sines and cosines. For H_n, $C < 0$ and $\eta_{AB}\dot{X}^A\dot{X}^B > 0$ because the curved metric is positive and consequently $\lambda > 0$ which means that the solution involves hyperbolic sines and cosines. Finally, in the limiting case $\lambda = 0$, we have the solution

$$X^A(s) = X^A(0) + \dot{X}^A(0)s. \tag{B.31}$$

The constancy of $\eta_{AB}X^A X^B$ requires that the initial velocity is light-like as possible in AdS_2.

Reference

[1] Bengtsson I 1998 *Anti-de Sitter Space* https://web.archive.org/web/20180319192552/http://www.fysik.su.se/~ingemar/Kurs.pdf

CPSIA information can be obtained
at www.ICGtesting.com
Printed in the USA
BVHW010458230421
605676BV00006B/66

9 780750 321853